Flying the Weather

FLYING THE WEATHER

The Story of Air Weather Reconnaissance

Otha C. Spencer

The Country Studio
Campbell, Texas

Library of Congress Catalog Card Number: 96-96887
ISBN 0-9653600-9-1

Published by
The Country Studio
Rt. 2, Box 54
Campbell, Texas 75422

To members of my family
who were with me through this adventure:
Billie, John, Mary, Joanne, Will and Gwendolyn

Hurricane Reconnaissance Flight
1944

Contents

Illustrations

Preface

The world's first weather flight was inaugurated in the mists of biblical history, unknown centuries ago, when Noah launched a dove from the *Ark* to determine weather conditions for landing:

> Then he sent forth a dove . . . but the dove found no rest . . . and she returned unto him into the ark, . . . yet another seven days . . . the dove came to him in the evening; . . . in her mouth was an olive leaf . . . so Noah knew that the waters were abated from off the earth.[1]

From this humble beginning in the world's prehistory, Captain Noah sent reconnaissance flights to help him decide when to unload his famous cargo.[2]

Through the years, weather has always been a concern to humankind. Starting with Aristotle, ancient seers, astronomers, witch doctors, and scientists have been trying to do something about it. An editorial in *The Hartford Courant*, in 1897, introduced a well-worn phrase into our lives, "Everybody talks about the weather, but nobody does anything about it."[3] "Now, after 2,500 years of fumbling, there is reason to believe that scientists can succeed, if given the opportunity, in a way that may affect the future of civilization even more drastically than the development of atomic power."[4]

We live in a thin atmosphere that surrounds our earth—a vast ocean of air, a symbiotic cocoon from whence come our life, health and the poetry of daily living. While the atmosphere has been compared to the thickness of a coat of paint on a baseball,[5] it not only supports life, but often generates a fury that brings death and the destruction of things we work for. The many moods of this ocean of air, from cool breezes, summer sunshine and life-giving rain, to the angry swirling winds of the tornado and hurricane, is what we call weather.

Throughout history, weather has brought good and evil. Genghis Khan, in the thirteenth century, used winter to cross the frozen Yellow River to defeat the Chin empires. In 1281, Kublai Khan failed in his invasion of Japan when Kamikaze, the "divine wind," wrecked his powerful Mongol fleet in Fukuoka Bay.

Christopher Columbus, on his earlier voyages to the Caribbean, learned to respect the "big winds of the New World." In 1502, on his last voyage, he saw signs of a hurricane threatening the Caribbean. Francisco Bobadilla, Spanish colonial governor of the West Indies, was about to depart for Spain, his ships laden with riches. As the storm approached, Columbus warned Bobadilla and suggested that he wait. He also asked the governor's permission to shelter his own ships in a river in Santo Domingo. The governor said, "no!" and sailed with his fleet. Columbus found a safe haven and rode out the storm, while the Spanish governor lost all ships but one and went down with his rich cargo.[6] To ignore weather is to invite disaster.

England's Sir Francis Drake defeated France's Philip II in 1588, when violent gales allowed Drake to outmaneuver and outshoot the French. Philip said, bitterly, "I sent them to fight the English, not storms."[7] Weather played a far greater role in the destruction of the French armada than did English gunnery.

The Greeks were the first to study the weather scientifically. In 350 B.C., Aristotle published four volumes which stood as the standard text on weather for two thousand years. In 1592, Galileo invented the thermometer, and, in 1642, Torricelli invented the barometer—the first major advances in weather measurement since Aristotle. During the Crimean War, in 1854, Emperor Napoleon's troops suffered a great loss from a Black Sea storm. He asked astronomer Urbain Leverrier to study the storms. Leverrier's recommendation led Napoleon to establish a system of national communications to study and track weather. Britain, Turkey and India quickly established weather services. However a recommendation to the United States Congress to establish a national weather service was not considered until 1870, when funds were appropriated. President Ulysses S. Grant, on 9 February of that year, signed the law that authorized the establishment of a weather service in the Army Signal Corps. Gen. Albert James Myer, who came from the Army Medical Corps, took command in November 1870 and put the weather service in operation. At his death, he was called "the most useful man of his age."[8]

In 1935, the weather service was transferred from the Signal Corps to the Army Air Corps so that the Army could be better prepared if war came. The Air Corps had a greater need for weather information than did the Signal Corps, and would also better understand the forecasting requirements of the civil airlines. In 1939, under the Chief of the Air Corps, weather commands were established in Alaska, Hawaii, Panama, Philippines and Puerto Rico, plus three weather regions in the continental United States. Headquarters and training commands were also established.

War dictated more organizational changes in 1943. The Air Weather Service operated under a chain of command from the Secretary of War, to the Army Chief of Staff, to the Commanding General of the Army Air Forces, through the Weather Division Assistant Chief of Air Staff, to the Army Air Force Weather Wing, Asheville, North Carolina. Under the Weather Wing, eight weather squadrons were located in the United States, and four squadrons were established outside the states, in Panama, the North Atlantic, Caribbean, Canada and North Alaska, and the South Atlantic Weather Wing, Air Transport Command (ATC), in Brazil. Other overseas weather squadrons were responsible to theater commanders.[9]

As World War II began, the use of weather reconnaissance was tentative, at best. Military planning seemed to be based on might against might, strategy against strategy. Weather, except as it hampered operations locally, was so transient that it was not considered a serious tactical consideration. A clear day at the takeoff field deceptively suggested that the bombing target was just as clear.

Although predicting weather movement through aerial reconnaissance had been practiced by the military in World War I and by the civil airlines during the buildup of air travel, World War II was the real stimulus for developing weather reconnaissance programs. Military leaders needed to predict weather for invasions, bombing raids, naval encounters, and movement of troops and aircraft, etc. Violent weather was also a civil threat, and the Army and the Navy were asked to provide aerial surveillance for use in warning civilian populations of dangerous weather. Most of the violence in weather comes from the oceans and when we realize that fifty percent of the U.S. population lives within fifty miles of an ocean,[10] warnings for hurricanes, Pacific fronts and Nor'easter ice storms become a life and death matter.

War is often the accelerator of progress. To fight a battle with too little or wrong weather information can be disastrous. In World War II, on 26 September 1942, an attempt to bomb German airfields with sixty-two B-17s was turned back because weather prevented Eagle Squadron 133, flying Spitfire IXs, from protecting the bombers. Through sloppy briefing, carelessness, poor weather forecasting, and bad luck, the Spitfires found a 100-knot tailwind rather than a forecast 35-knot headwind. Twelve young pilots and twelve new Spitfires were lost.[11]

Modern commanders often forget the lessons of history and ignore the violence of weather. On 17 December 1944, Adm. William F. "Bull" Halsey drove his Pacific fleet into a typhoon east of Luzon. The fleet fought the winds and towering waves for two days and were helpless to get out of the storm. Three destroyers, with 790 men and 146 combat aircraft, were destroyed. Aircraft carriers, cruisers and smaller ships were badly damaged. Halsey's typhoon was called the "most crippling blow to the Third Fleet than anything less than a major action."[12] Yet, on 4-5 June 1945, Admiral Halsey made the same tragic mistake and steered his fleet into another typhoon. Thirty-six ships were damaged, including the carrier *Hornet*, whose forward deck was caved in, and the heavy cruiser *Pittsburgh* lost over one hundred feet of her bow.[13] As a direct result of Admiral Halsey's encounter with the typhoons in 1944 and 1945, the first Navy weather reconnaissance units were formed in 1946.[14]

In 1942, the idea of dedicated weather reconnaissance with specially-equipped aircraft was conceived, and, following this idea, a new weather program was inaugurated in 1943.[15] A squadron of B-25D weather planes was put in operation, patrolling the North Atlantic ferrying lanes, where air traffic was heavy with lend-lease aircraft being flown to Europe. This marked the beginning of organized aircraft weather reconnaissance with the sole mission of providing weather data for military and civil use.

To have an accurate weather forecast, observations from thousands of points around the world are being taken constantly and evaluated by meteorologists before a forecast can be issued. These observations come from aircraft, ships at sea, ground weather stations, weather volunteers and the weather services of friendly, cooperating nations over the world. Even so, in many areas, such as the untracked expanses of the great oceans, the Arctic, the Antarctic, and in undeveloped countries, there are no observations and forecasters are severely handicapped. To fill these gaps, a world-wide network of flying reconnaissance routes was organized with trained meteorologists taking in-flight weather observations. These were called "synoptic" flights, made daily on long dull, fixed-track missions over isolated parts of the world, especially the oceans.

Synoptic reconnaissance flights formed the real backbone of weather observation. More glamorous missions such as hurricane and typhoon recon, chasing hot clouds of atomic radiation, or spotting weather for space probe blastoffs or splash downs were weather missions that made the headlines of the world, but, basic weather recon was synoptic.

In planning the invasion of Normandy, weather was vital to success. The Allies, using a new type of forecasting, were successful in determining the weather needed. The Germans had completely misread the weather and failed to foresee the D-Day landing. So, on 5 June, Admiral Theodore Krancke, Navy Group commander, began an inspection trip to Bordeaux because, "according to weather reports, there seemed to be no chance of an invasion within the night of 6 June."[16] But history records that there was an invasion, and it was successful, in part, because weather recon aircraft reported actual weather conditions when the troops landed and the bombers bombed.

I consider myself fortunate to have been a part of early weather reconnaissance. This history is written from actual experience. In the spring of 1944, I began as a recon pilot with the 2nd Weather Reconnaissance Squadron, flying route reconnaissance over the frozen North Atlantic from Labrador to Greenland, Iceland and Scotland. Later, in the summer of 1944, I was pilot on one of four air crews, flying B-25Ds, assigned to the Ninth Weather Squadron, Morrison Field, Florida, "for the purpose of reconnaissance of tropical storms and hurricanes."[17] However, as a history, this is not my story—it is not a personal memoir. Weather recon was continued long after I left active duty. But I continued to "fly" the missions and "feel the experience" of weather flying.

Our crews flew hurricane reconnaissance during the 1944 storm season. Col. John K. Arnold, regional control officer of the Ninth Weather Region, after the 1944 storms, called hurricane reconnaissance an "entirely new operation, both from the standpoint of flying and of weather reporting and forecasting."[18] This operation was also the turning point for weather reconnaissance as exclusively a military tool, to reconnaissance for the benefit of the nation as a whole—accurate weather reports and warnings of approaching storms were as important to civilians as to the military. Although this history focuses on the work of the of Army and Air Force in weather recon, we cannot ignore the Navy, whose planes and crews often shared missions with Army crews. The Navy patrolled coastal areas for the protection of ships at sea, flying PB4Y-1 *Liberators* and PB4Y-2 *Privateers*. Later, the WC-121 *Super Constellation*, with its huge "radom" (radar dome), contributed to locating and tracking hurricanes.[19] The Navy history of weather recon is yet to be written.

In 1949, a new danger blended invisibly into the atmosphere of the world. An Air Force weather reconnaissance plane, flying at 18,000 feet from Japan to Alaska, detected signs of intense radioactivity over the North Pacific—twenty times normal. David Lilienthal, administrator to President Truman, described the newly discovered radioactive cloud as "a whole box of trouble."[20] Three days later, after the evidence had been thoroughly analyzed, President Truman issued the statement, "We have evidence that within recent weeks an atomic explosion occurred in the USSR."[21] Since that time, weather reconnaissance planes and crews have been responsible for flying into the "hot clouds" of atomic explosions to detect the source and movement of radioactive particles in the atmosphere.

In interviews with hundreds of weather crews, unusual and strange stories were told about weather recon as an entirely new frontier of flight. Flying techniques were learned that made military and civilian air travel safer, and the limits of crews and their aircraft were reached and extended. New discoveries seemed almost limitless. B-29 bombers and weather recon planes discovered the jet stream over Japan in World War II; weather crews learned how to control Britain's black fog that shut down airports; weather navigators learned how to make their way through the unexplored Arctic regions where the magnetic compass was useless; pilots conquered the hazards of "flying in the milk" over the Greenland ice cap and the blinding haze of oceans where earth and sky blend in a Zen-like realm of no up or down.

This book is an oral history, told by those who flew weather, with the following goals:

(1) To outline an accurate historical saga of weather reconnaissance from 1916 until the present, in all parts of the world;
(2) To tell the reconnaissance story through the experiences of the men and women who flew the weather missions;
(3) To allow readers to feel the "weather recon experience"—the brutal drama of flying into typhoons and hurricanes; the surrealistic feeling of flying to the North Pole; and the fear in being surrounded by deadly particles, floating in the atomic hot clouds;
(4) To explain the special relationship of flight crews and their aircraft as they battled and won out over the worst that nature could put forth—a special blend of mind, spirit, muscle, aluminum, design, power and luck.
(5) To record aircraft weather reconnaissance's eventual submission to "that great electronic 'aircraft' in the sky" called the satellite.

In spite of its success in military operations and its potential for protecting lives, weather reconnaissance suffered a low priority in national planning, and the story of aerial weather surveillance remains a dark chapter in military history. Weather crews almost always flew surplus, cast-off bomber or cargo aircraft, most of which were battle worn and pitifully suited for reconnaissance. There was never an aircraft designed specifically for weather reconnaissance to reach regular operational use. Constant budget restraints and the rigidity of military leadership made it difficult to consider aircraft beyond pure tactical and defensive use. In conducting weather reconnaissance for civilian use, the military was asked to do what the Department of Commerce should have been doing. The smallest of all wartime reconnaissance programs, and the last to develop, was weather reconnaissance.[22] Military historians Wesley Frank Craven and James Lea Cate wrote, ". . . their services . . . were taken for granted and little thought was given to the near-miracle they represented."[23]

Today, as orbiting satellites look down on the weather, air-borne reconnaissance has been reduced to one U.S. Air Force Reserve (USAFR) squadron, and the research crews of the National Oceanic and Atmospheric Administration/Aircraft Operations Center (NOAA/AOC). The reserve crews, known as "*Hurricane Hunters,*" make up the 53rd Weather Reconnaissance Squadron, stationed at Keesler Air Force Base, Biloxi, Mississippi, flying aging WC-130s, modified four-engine, turboprop cargo aircraft. NOAA flies Lockheed WP-3 "*Orions,*" another turboprop, and is spending millions on the purchase and modification of new jet aircraft.[24] NOAA crews are stationed at MacDill Air Force Base, Tampa, Florida. The 53rd is authorized for twenty air crews for reconnaissance, and NOAA maintains two *Orion* "flying laboratories."[25]

Even with present orbiting satellite systems and limited aircraft recon, storm warning centers are not able to provide the information needed for completely adequate violent weather warnings. Dr. Robert C. Sheets, former director of the National Hurricane Center, Coral Gables, Florida, speaking specifically of hurricanes, said at a Congressional hearing on violent weather, "We clearly need more quantitative data over the open ocean. . . . A hundred years of hurricane records . . . are not going to change, . . . they are going to be there. But I like this message from a church marquee in Harlingen, Texas: 'I cannot direct the wind, but I can adjust my sails.'"[26]

The machines the crewmen flew were as much a part of weather recon as the men and women themselves. No writer worthy of the pen has ever written about the airplane without searching for all the poetry at hand to describe this obsessive Great White Whale that frees us from our earthbound existence. Writer and flyer alike know that "airplane" is a feeble, puny name for this Lord of the Skies—aeroplane is better; Ernest K. Gann called it "The High and Mighty;" to Antoine de Saint-Exupéry it was his passion, his home and his coffin. The dreams of Leonardo da Vinci and the fantasies of Jules Verne fired the imagination of Orville and Wilbur Wright, who put wire, sticks, sprocket wheels and guts together to teach man and metal to fly. It's a jaded story now and only the truly romantic still peer into the sky and marvel at "the strange high singing of the aeroplane overhead. . . ."[27] To the writer, who, as a flyer in more youthful days, shepherded those monsters through nature's wilderness of clouds and stars, there rests a double burden to remind his reader that flight is the truly Grand Miracle of the twentieth century, and being a part of it was an experience of the soul. To fly them is a soft dream and, often a violent nightmare, but every plane is the best that any pilot ever flew because pilot and plane become one—a being that never before existed—as they make their way in a wind that never before blew and stealing into places where no one had ever flown.

The men and women who flew weather adopted a cavalier attitude toward fighting nature's extreme moods. They were called "*Hurricane Hunters*," "*Typhoon Goons*," "*Arctic Blue Nosers*," or simply "*crewdogs*." I have had hundreds of letters, conversations and interviews with weather officers, flight commanders and other crew members who battled nature's fury and won. They spoke proudly of their work in the air and of those on the ground who kept the planes flying—each had equally important stories to tell. They spoke quietly of those who lost their battle and those who cracked during the fight.

History as a record may be "true," but history as a memory may be "more true." As years pass—dedicated weather recon began over fifty years ago—memory has intensified, and the events of history may have become "more real." I believe that the stories of the men and women who flew the missions speak of the reality that existed at the time, although we may have been too young, too serious and too far away from home to realize the pure drama of our flights. To review history from those who made that history helps to preserve many an experience that might have been lost in a strict "place and date" record. My history professor always told his classes, "History is the part of our lives that we remember, and if not true, it should have been and perhaps actually was."[28]

This is a book for the weather lover, for pilots and flying men and women who love the lore of the air and want to know more about weather, and this is a story for historians interested in the fringe areas of the larger picture of war and flying. I write this history from the sheer love of weather and flying and I share with the poet,

> . . . every sky has its beauty, and storms which whip the blood do but make it pulse more vigorously.[29]

But, most important of all, I write for the crews in the air and on the ground who ventured into the beauty and violence of weather and tried to relate their experiences as a living history of those who probe nature to appreciate, to understand and, yes, to warn.

Acknowledgments

The author is the one who writes the book and whose name is proudly displayed on the bookjacket. However, just as the pilot of a weather reconnaissance aircraft could never hope to get into the air without the support of a crew, the author could never have written this book without the encouragement and support of hundreds of weather recon veterans, family and friends. The idea for this history came from David Magilavy, president of the Air Weather Reconnaissance Association. After the idea, a few telephone calls brought the support of Robert Mann, president of the "Pacific Air Weather Squadrons Association," a group of veterans whose aircraft covered more than sixty-three million square miles of the Pacific Ocean. Then Retired Major General John W. Collens, president of the Air Weather Association, gave his endorsement. These men and members of their organizations sent a virtual flood of information, personal experiences, tapes, photographs and military histories. At the reunion of the Air Weather Association at Tucson, I received so much research material I had to buy an extra large flight bag to carry it.

There followed three years of personally rewarding research and writing—more than a thousand letters were written and hundreds of telephone calls were made. My own weather recon experience came back to me and I met fellow recon veterans from all parts of the world. I wish I could acknowledge all who have helped with this book. Most are mentioned in the retelling of their experiences. To all I say, "thanks,"—this is your history.

To my family the book is dedicated—fifty-two years ago, in 1944, my wife shared the experience of my first major hurricane flight and she must now share this history. To Dr. Fred Tarpley, historian and word specialist, goes my appreciation for reading the manuscript and I still have a few commas left over from his editing. David Magilavy kept the writing meteorologically correct. Vivian Freeman set the book in type and designed the pages. Dorothy London watched as it went through the presses. I thank them all.

With the help of this support group, I hope our mission is a success. "No man is an island."

Flying the Weather

Prologue

Flying weather reconnaissance is different from any other mission Air Force men and women were called on to fly. Weather was the invisible enemy—sometimes quite beautiful and guileless—sometimes stormy, ugly and deadly. Even the most experienced crew members often panicked when their plane was wrapped in clouds, with turbulence, winds and lightning threatening destruction of crew and aircraft. The following stories will let you feel the experience of weather reconnaissance.

The earliest recon was tactical: along the route and over the target. The 107th Tactical Reconnaissance Squadron of the Eighth Air Support Command was a group of dare-devil, hot-shot weather pilots. The 107th, called "Mercury," was a squadron of ten retrained B-26 bomber pilots with five P-51B Mustangs, stationed at Middle Wallop, England. Their job was to penetrate German territory, scout the Eight's bombing targets and report weather along the way.

Mercury was designated the Ninth Weather Reconnaissance (Provisional) Squadron, under the command of Maj. Maxwell W. Roman, pilot-meteorologist. "Provisional" status created a normal military SNAFU.[1] All personnel of the Ninth were on "detached service,"[2] "borrowed" from other squadrons throughout England—pilots even brought their own aircraft. The squadron was practically unknown to its "on paper" commanders, and pilots and ground crewmembers were consistently overlooked for promotions or decorations. Twenty-five officers were lieutenants, the line chief was a sergeant, and crew chiefs were privates. They called themselves "Orphans of the Storm." They are not recognized by any weather recon organization, and formal history makes little mention of the Orphans.

Assigned to the Ninth Bomber Wing, they were on call to fly weather missions for the Ninth's bomber strikes. Facing death on every mission, the close-knit "Orphans" developed a "street-gang" arrogance, a special *esprit de corps* with a distinctive RAF flavor. British uniforms were mixed with U.S. GI issue; missions were "trips,"—a planned mission was a "do," and reconnaissance was "recce"; a character was a "type," *(he's a good type)*; women and mademoiselles were "baby dolls"; the possessive "my" was "me," *(me arse, me shoes, me friggin sack)* and their aircraft were "kites."[3] The "Orphans" were often compared to Claire Chennault's famed "Flying Tigers"—the American Volunteer Group in China. The Orphans and the Tigers treasured their bastard status and unbelievable flight accomplishments—everyone loved them—no one claimed them.

In early 1944, the Orphans were assigned to patrol the Normandy invasion area and, between 4 January and 6 June, they flew 275 missions (550 sorties), preparing for D-Day.[4] During the Normandy invasion, the recce planes flew twenty missions up and down the beaches. Before darkness fell on 6 June, Orville Scott, from Ola, Arkansas, with 141 weather missions, and his fellow cloud chasers, had made ten flights over Hitler's West Wall, spotting weather for bombers, fighters and fighter bombers of the Allied air fleet.[5] From 5 June until December 1944, the Orphans flew another 1,362 combat recon sorties.[6]

In the single-engine P-51 Mustangs, pilots would fly two-plane missions—one plane for recon and one for fighter protection. Their roles often changed, even on a mission. Recce planes were over the target before the B-26 bombers took off, so that if the weather was "no," the bombing mission could be canceled. Pilots sent weather messages, called "flash reports," in the clear to "parade" since there were too many errors in coded messages, and there was no time to encode and decode. Only the target was in code. A flash report was to the point: "Damn poor climate today. Good thing we buzzed over. Saved those Marauder boys a rough afternoon . . . and one helluva lot of gasoline."[7] If the mission was "do," a second recce plane was over the target as the bombers crossed into enemy territory to make sure the weather had not dropped below minimums. Weather planes flew erratic courses and often checked several targets to confuse the Germans.[8]

Flying weather recon in a P-51 required special skill. The aircraft was fast and temperamental and, with only one engine, there was no safety margin in an emergency. Pilots "invented" daring techniques to fly weather and learned survival skills that have been used by other recon crews throughout the history of weather flight. When hopeless weather grounded other planes, the Orphans continued daily missions. Kansas-born Lt. George Brooks, veteran of eighty-four missions, flying his *Kansas Aggie*, explained how the Orphans flew out of fields with zero-zero conditions: ". . . [In the plane] I buckled in and started up. . . . Sergeant Nance, the crew chief, with his hand on the wing, lead me to the center of the runway. I carefully lined up with the vectors of the runway using the magnetic compass, then minutely set the gyro compass . . . steadily applied throttle, not letting that gyro vary at all . . . lift off! wheels up! and on my way!"

It is easier to take off in soupy weather than to land, but any pilot will tell you that landing is the most important part of any mission and ingenious ways had to be worked out to get on the ground safely. Laon Field, at Middle Wallop, was built on a 400-foot pile of rocks, with a 300-foot cathedral spire and hills on the south and east. Capt. Warren Harding developed a system for landing in bad weather, which pilots practiced when the weather was good. When planes came in from a mission, "The pilot would call the DF station, get a steer and come across the field at 180° at 1000 feet . . . hold for one minute, turn 90° to the right for 15 or 20 seconds, and take a heading of 340,° making a letdown in the turn." There was a valley southwest of the field, and Captain Harding's procedure allowed a safe letdown into the valley. "Many of us owe our lives to Sergeant Dixon and his crew in the DF station," Harding wrote.[9]

Pilots were political enemies on the ground; in the air, they were professionals and developed a deep respect for the other's skills. On one mission, Lts. Andy Clark and George Brooks saw a lone Messerschmitt BF-109 some 8,000 feet below them. The German pilot saw them, dropped his belly tank, and squared off to meet their attack. The pilots soon realized they had each found a real pro. For screaming minutes, it was Mustang-Messerschmitt-Mustang with neither able to gain an advantage. Finally, after pulling all the Gs[10] they could, the German plane snapped into a spin and disappeared—they claimed one Messerschmitt wing tip. Brooks conceded, "I know damned well he recovered . . . he was too good. I'd treat him to one helluva supper if we would run into each other now."[11]

Victory in Europe brought the end to the Ninth Weather Reconnaissance Squadron (Provisional). The Orphans had flown weather missions for the Normandy invasion, were the first to report V-1 launching platforms, V-2 rocket contrails and the new German jets. Their recce missions saved the lives of countless Allied airmen that might have been wasted seeking targets that could not be found. On V-E day the Ninth "faded back into the mists that for so long had been their home . . . to become orphans of history."[12]

Synoptic reconnaissance included the hostile world of bitter cold. Cold weather operation presented new problems, as weather crews found in the North Atlantic, and on flights to the North Pole. The extreme cold tested crews and aircraft to their limits, and new and unknown weather phenomena were a constant challenge. Kent Zimmerman, San Antonio, Texas, aircraft commander for the 514th WRS, describes flight and ground operations in Arctic weather: "I came into Eielson AFB (Alaska) on a flight from Travis AAF (California) when the temperature was forty degrees below zero. The runways had been plowed and snow banks on either side were ten feet high—pine trees were used to mark the edges of the runway which was covered with about three inches of ice crystals. As I landed and reversed the propellers to brake, the runway disappeared in a blinding snow storm. I immediately took the props out of reverse and rolled out of the cloud of ice crystals that had blown up in front of the aircraft, but I had been rolling down the runway completely blind, for about half of its length, . . . using the nose wheel to steer. The ice crystals were so light that a small wind or an aircraft taking off would close the runway for ten to thirty minutes.

". . . I parked on the ramp for service. Gasoline was so cold, the smallest drop would frostbite your skin, and you didn't dare touch the aircraft metal. You could spit in a bucket of gas and the spittle would freeze instantly; the oil was so cold it would come out of the nozzle like soft tar. If you didn't dilute the oil in the engines before shut down, they could never be started again. Twelve hours before a mission, the aircraft was towed into a warm hanger, and, if you had overfilled the tanks, the gas would expand and run out the vents on the hanger floor. After briefing and flight clearance, the crew, with lunches and flight gear on board and the aircraft still in the hanger, would pre-flight using auxiliary power. The plane would be towed outside and you had ten minutes to get the engines started and temperatures up to max takeoff power. Down the runway for takeoff—the air is dense, takeoff roll is short—and into the mission."[13]

Lt. Col. Robert Ruark, pilot on a *Ptarmigan* mission, reported similar problems: "In the intense cold, the RB-29s were kept inside hangers so that ground crews could work. The flight crew would pre-flight the plane in the hanger and a Cletrac[14] would tow the aircraft out. Once outside, no time was wasted in getting the engines started. Our biggest fear was having a false start. If the engine fired one time and quit, the spark plugs would freeze. We then had to stick a heater hose into the engine intake so that hot air would thaw and dry out the plugs. Other problems included frozen and ruptured oil lines, trying to keep air in the shock struts—everyone had to wear gloves to keep fingers from sticking to the freezing metal. As we taxied to takeoff in the icy winds, the airplane would creak and groan as the metal contracted, and we were happy to be airborne so that leaking fuel would evaporate. Once in the air, inside the plane, the temperature would drop below zero and we wore heavy woolen underwear, shirt and pants, sweater, and a wool flying suit with fleece-lined pants and parka. To protect our feet, we wore several pairs of socks, felt inner liners, inner socks and mukluks, and our hands were protected by three pairs of gloves, nylon, wool and leather."[15]

While extreme cold was a problem, the real challenge in the polar region was navigation—the simple act of going from here to there. There was no LORAN (**Long Range Navigation**), and each flight had as many as three navigators, all making star shots during darkness and sun lines during the day. Navigation was most difficult when the Arctic was in twilight and neither star or sun shots were possible. The time-honored magnetic compass was useless due to large variation errors,[16] and unknown wind drift made accurate DR navigation impossible. Navigators were frustrated, "When you know damned well that you're headed north and the magnetic compass is wildly pointing in odd directions, you're not sure if you're coming or going."[17] In the fall and spring, navigators depended on positions of the moon and the brighter planets—Venus, Jupiter or Saturn—and during those seasons, there was only a forty-minute "window" when a flight must go or be canceled. Strange, unearthly, and "curiouser and curiouser" things happened. Crews often had the experience of seeing two sunrises and two sunsets on one mission. The sun came up on the flight northward, set when they were almost at the Pole, rose again when they were headed south, and set before they landed. [18] The airmen were strangers in a strange land, and one crewmember even remembered a Lewis Carroll poem: "The sun was shining on the sea, / Shining with all his might: / This was odd, because it was / The middle of the night."[19]

Hurricanes and typhoons have roamed the earth like atmospheric giants, killing and destroying since time began. Radio news commentator, Edward R. Murrow, after a hurricane flight through *Edna*, in 1954, described the experience: ". . . many hours of boredom, interspersed with a few minutes of stark terror. . . . The eye of a hurricane is an excellent place to reflect upon the puniness of man and his works. If an adequate definition of humility is ever written, it's likely to be done in the eye of a hurricane."[20]

Capt. Robert E. Guthland, meteorologist for the 54th Weather Reconnaissance Squadron (WRS), from Warrensburg, Missouri, flew a night penetration of the Pacific typhoon *Dinah*, in 1958. Guthland and his crew were flying the WB-50, one of the largest and toughest of all hurricane recon aircraft.

The mission was to make a sunset fix on *Dinah* moving westward toward China. Taiwan and the Philippines were under the influence of this storm, and the mountains of Taiwan made radio fixes after sunset impossible. After takeoff from Clark Field, (Philippines), in mid-afternoon, the crew fought solid instrument conditions during the entire flight to the storm. Penetration was made by Doppler radar with moderate turbulence and winds of 100 knots (115 MPH) at flight level. "After penetration of the eye," Guthland explained, "difficulties in navigation required us to remain in the 15-mile diameter eye until an exact fix could be established. We were happy about this since the delay in the eye was the first time we had been out of the clouds since takeoff. A hurricane, from the inside, is a rare, unearthly sight, seen by only a few people. The late afternoon sun's rays glistened through the anvil tops of the ominous wall clouds—an awe-inspiring sight. This wall cloud has been described as 'that canyon of stillness in the heart of every mature hurricane.'"[21]

As Guthland and his crew left the eye to head back to Clark, they were ordered to return to *Dinah* for a midnight fix, since the storm had made a turn to the north and was headed toward Okinawa. This new position required another fix and new observation reports. For five hours the big WB-50 wallowed in the fringes of the storm and the crew watched the eye by radar 150 miles away—the intensity of the storm lit the radar scope brilliantly. Guthland writes, "As midnight neared, we started toward the eye and turbulence intensified. I suggested that we take a radar fix and head home. Everyone cheered. However, that was impossible. Navigational aids had been so disrupted by turbulence and lightning that we had no idea where we were—we only knew our position from the storm by radar. So we had to get back in the eye, not only to fix the storm, but to 'fix ourselves.' Once within the typhoon eye, radio and Loran reception were abnormally good."

The next fifty miles offered more than any crews' share of thrills. The pilots had all they could do to keep the aircraft in level flight at 9,800 pressure feet. Lightning was flashing through the sky and the plane's air speed varied from 300 to less than 150 knots—one minute flying at 14,000 feet, the next minute at 8,000, and the big plane was being carried to the right of the eye by terrific crosswinds. "We tried to turn left but we were being tossed around the eye of the storm. We fought updrafts and downdrafts— lightning hit the plane and bounced around the cockpit,—the entire aircraft was bathed in St. Elmo's fire. Five miles from the typhoon center, five fire-warning lights came on and Number 3 engine began to backfire— the radio operator called that his transmitter had broken loose and was flying around in his compartment.

"We broke into the eye at 12,000 feet and it was only three miles wide—we had been pushed 180 degrees from our starting point. After a dive to the 9,800-foot dropsonde ejection altitude, we hit the wall cloud again. It seemed as if a giant hand stopped the aircraft in mid air. The WB-50 shook, every bit of metal seemed to screech in anguish, and we were literally tossed out of the storm through the wall cloud. In that one minute in the eye, the navigator found our position, the dropsonde operator made his drop, and we headed back toward Clark. We landed with bare minimum conditions after a flight of eleven hours of solid instrument time, excluding the time in the eye. None of the crew will every forget our "midnight affair" with *Dinah*." [22]

As atomic power and weapons were being developed around the world, "hot" radioactive clouds began drifting through the air currents of the earth and weather recon had another duty. In the spring of 1951, as the cold war was turning hot, a series of atomic tests were scheduled for three islands on Eniwetok Atoll in the South Pacific. These tests were called *Greenhouse*, a joint military-civilian project with the four military services and the Atomic Energy Commission (AEC). [23] Weather reconnaissance crews were assigned to track atomic clouds from the tests and determine their movement and intensity from ground zero.

For the *Greenhouse* tests, the entire 57th Weather Reconnaissance Squadron (WRS) went to the Marshall Islands for atmospheric sampling. This introduced a completely new mission, there was much apprehension, and crews reported unusual experiences. Test bombs were being dropped, and on the fourth detonation, on Entebbe Island, aircraft commander Col. Eugene Wernette, Fort Worth, Texas, reports his experience:

"On 25 May 1951, our crew departed Kwajalein and climbed to 29,000 feet and set up an orbit outside Entebbe, about five miles from ground zero. Above us was a flight of drone B-17s, one of which went out of control and spiraled down through the other aircraft. At zero minus fifteen, we put on dark glasses, folded arms across our chest, closed our eyes and looked down into the cockpit. At zero, a white light flashed—so bright it made my arm and hand a skeleton. That intense white brightness just cannot be described and I will never forget my feelings at the time. There was absolute quiet in the plane—not one of the crew spoke a word anticipating the shock wave which rocked the aircraft sharply.

"At zero plus ten, we were allowed to look without glasses. By that time the intense light was gone and was replaced by a giant fireball and a boiling mass of clouds. When the clouds reached our altitude, Mitchell, the radiology officer on board directed us into it for samples. He immediately said, 'Let's get on the ground fast. . . . So, we departed the cloud and, in the clear, headed down at red-line speed, and landed at Eniwetok.

There we were met by David Magilavy, the radiology safety officer, and his team. We took off all of our clothes and went into the shower tent. After several scrubbings and checks, we received new flying clothes. The aircraft was washed and we returned to Kwajalein. About ten years ago, I learned our exposure was 4.9 rems.[24] At the time I considered it a hazardous flight and I still feel that way."[25]

1

The Beginnings of Weather Reconnaissance
1916-1944

The greatest military invasion of all time was set for 5 June 1944. More than two million Allied troops, 5,000 fighting ships, 4,000 landing craft and 11,000 aircraft were poised in England, ready to cross the English Channel and storm the beaches of Normandy. Sixteen million tons of munitions and supplies were stockpiled. With a critical eye on the weather and the extreme pressure of his responsibilities, Supreme Commander Gen. Dwight Eisenhower agonized over his decision to invade Europe. Prime Minister Winston Churchill protested against the futile slaughter,[1] "When I think of the Beaches of Normandy choked with the flower of American and British youth . . . and when in my mind's eye, I see the tides running red with blood, I have my doubts. I have my doubts."[2]

The Normandy invasion was code-named *Operation Overlord*—commonly called D-Day.[3] Allied bombers had been pounding Normandy military installations for weeks to prevent a German build up. Selection of the actual day for the assault depended on weather forecasts, with 5, 6 and 7 June being considered. However, weather forecasters were predicting stormy weather for 5 June, to be followed by a thirty-six-hour period of good weather before another storm from the North Sea would move in.[4] General Eisenhower, in his book, *Crusade in Europe*, wrote,

> If none of the three days should prove satisfactory from the standpoint of weather, [the] consequences would be almost terrifying to contemplate. Secrecy would be lost. . . . assault troops would be unloaded and crowded back into assembly areas . . . complicated movement tables

would be scrapped . . . morale would drop . . . a wait of at least fourteen days, possibly twenty-eight . . . a sort of suspended animation
involving more than 2,000,000 men![5]

What kind of weather did General Eisenhower want? His needs were
clear: "No high winds to raise swells that might batter ships in crossing
the Channel; ceilings no lower than 11,000 feet for heavy bombers, less
for medium bombers; 2,500 feet for cargo and troop carriers with visibility of three miles; 1,000-foot ceilings for fighters and no more than 35 MPH
winds for paratroopers. . . ." When storms were predicted on 5 June, Eisenhower, delayed the invasion until 6 June and troops spent an anxious twenty-
four hours waiting for the weather to clear.

Allied weather advisors from the U.S. and Britain had differing opin-
ions, and a bit of international tension developed. Dr. James Martin Stagg,
British civilian and internationally respected meteorologist, was chief weather
officer to General Eisenhower. Col. Donald Yates, senior USAAF weather
officer in the European theater, was his deputy. Stagg's weather analysis
differed from the forecast of senior U.S. forecaster, Lt. Col. Irving P. Krick.[6]
British forecasters, using the traditional method of weather prediction, warned
against 4, 5 and 6 June because of unstable conditions over the English
Channel, and rain was approaching the British Isles and Europe. However,
the U.S. weather staff, using a new method of long-range forecasting, developed in 1943 at the California Institute of Technology, insisted that a
high pressure system over the Azores would push the front back over
Europe, and 6 June would be a good day. General Eisenhower accepted the
U.S. prediction. Still, on the sixth, light stormy weather persisted, and, while
not perfect, the weather was operational,[7] with overcast skies, rain, high
wind and choppy seas. The invasion was on, and Gen. Sir Bernard Montgomery's infantry and armored troops hit the beaches at 0630. Six hours
earlier, shortly after midnight, paratroopers had been dropped behind the
lines to capture landing fields, destroy bridges and disrupt rail lines. Gen.
Omar Bradley blamed weather for his troops being scattered "far and wide" of
their objectives.[8]

Most people assumed that the weather was not a factor in the invasion.
The truth is that weather was extremely significant during the crucial first
hours, and the predicted good weather did not come until late in the day
of 6 June.[9] While there may always be a debate that 7 June might have been
the best day for the invasion, the decisions of the German high command
seems to indicate differently. The Germans, pioneers of air weather reconnaissance, were quite accurate in their forecasts. But, as the anticipated
"window" for the invasion approached, German Maj. Gen. Richard Haber-
mehl, of the Reich's Office of Weather Service, was convinced that the invasion would not be attempted for another two weeks since he believed that

the winds on 4-6 June would be too strong for an Allied landing. So, on 5 June, Adm. Theodore Krancke, German Navy Group West commander, began an inspection trip to Bordeaux, France, because, "according to the weather reports available, there seems to be hardly any chance of an invasion within the night of 5 or 6 June."[10] Lieutenant Colonel Krick, wrote later, ". . . Because the Germans listened to bad advice they were caught by surprise and poorly prepared for our attack. The victory of our meteorology was complete."[11]

General Eisenhower admitted that if the invasion had not taken place on 6 June, the Germans might have mounted a stronger defense and possibly could have changed the outcome of the war and the course of history.[12] Military historian John Fuller calls this forecast "the most famous and important ever made," and eminent forecaster Sverre Petterssen described it as "meteorology's finest hour."[13] Norwegian Petterssen was one of the developers of the air mass frontal theory.[14] Part of the success of weather prediction for D-Day can be attributed to new ways of forecasting. Synoptic data for making forecasts, especially for weather moving in from open ocean or the North Atlantic, was gathered by weather reconnaissance aircraft patrolling the ocean. In addition to the tactical recon of the 107th Tactical Reconnaissance Squadron, calling themselves, "Orphans of the Storm" (See Prologue), the British weather service pioneered overwater synoptic weather reconnaissance.

Long before D-Day, Britons were fighting for their existence. In March 1941, the Lend-Lease Act was passed by the U.S. Congress, allowing the United States to send weapons, food and military equipment to any nation fighting the Axis. President Franklin D. Roosevelt called the lend-lease "helping to put out the fire in your neighbor's house before your own house caught fire and burned down."[15] While most lend-lease supplies went by ship, the fastest and most direct route for ferrying aircraft from the United States to England was across the weather-torn North Atlantic. The problems along this route were severe weather flown by inexperienced pilots. At the beginning, most ferry aircraft were flown into Langford Lodge, in Northern Ireland, where Lockheed Overseas Corporation had set up a base before military personnel arrived. At Langford Lodge, Lockheed, in addition to repairing and servicing battle-scarred aircraft, modified new planes, especially B-17 *Flying Fortresses* and P-38 *Lightning's*. Ferry pilots, stationed at Maghaberry, Ireland, came to Langford Lodge to get their planes. The pilots were called "Maghaberry Hot Shots," and they flew in weather when "even the birds were walking." The Hot Shots did their own weather recon, and are due a lot of credit for their daring; otherwise, British weather would have stopped the flow of bombers and fighters to combat pilots.[16]

Slow moving glaciers flowed like rivers from the mountains of Greenland to the sea where they broke into icebergs. This hostile terrain greeted the North Atlantic recon flights and was photographed from the western side of Greenland near Bluie West-8. *Photograph courtesy of W. A. Merrill*

About this time, in the early summer of 1941, before the war had threatened the United States, Col. Elliott Roosevelt, son of the President, had surveyed locations for a series of weather stations in northern Quebec and on Baffin Island, above the Arctic Circle.[17] Two days after Pearl Harbor, Maj. Don Yates and Maj. Arthur Merewether were sent to Labrador to set up the weather stations to support ferrying aircraft across the Atlantic to Great Britain. Major Yates, flying a B-18, was the first pilot to land at Goose Bay, Labrador, on a makeshift runway of hard-packed snow covering stumps of recently cut trees. Yates and Merewether organized the Crystal weather observing stations, Crystal I at Fort Chimo, northern Quebec; Crystal II at Frobisher Bay, Baffin Island; and Crystal III on Padloping Island, off the northeast coast of Baffin. They also established the Bluie West and Bluie East (BW-1, BW-8 and BE-2) air bases along the western, southern tip and eastern coasts of Greenland.[18] This network of stations provided weather support and landing bases for the ferrying route from Goose Bay to Greenland, Iceland and Scotland. Yates also proposed a ferrying route for American fighter planes from Alaska to Moscow.[19]

Early in 1942, as the Japanese were becoming a menace in the Orient, one of the Air Force's Weather Service's staff officers, Lt. Col. Torgils Wold, first commander of the 2nd Weather Squadron, in preparation for the war against Japan, had flown to New Delhi, India, to establish a weather region while the Tenth Weather Squadron unit could be manned, equipped and sent to India and China. The first commander of the Tenth was Maj. William E. Marling, and he was subsequently replaced by Col. Richard E. Ellsworth. On his first flight to India, Colonel Wold was piloting a B-24 carrying 21st Aviation Engineers to build airfields in the China-India-Burma theater. Jack C. Callaway, Arlington, Texas, was one of the engineers on the flight. Building airfields in India and the vast primitive areas of China is a remarkable story in itself, but has little to do with weather reconnaissance except for providing the airbases from which weather planes could fly.

The war was young and airmen were young and inexperienced. Callaway reported a potential mishap on the flight from Brazil to Africa: "Wold's navigator was fresh out of a ninety-day navigation school. After departing from Natal, they flew and flew and did not make landfall. Col. Wold called the navigator to the cockpit and asked why there was no land in sight. The very young navigator, tears streaming down his cheeks, said, 'Colonel, sir, I have missed Africa!' Wold knew what had happened. When the navigator had made a double-drift reading to find the wind, he had applied the drift in the wrong direction, making the correction doubly wrong. Wold turned sharply to the left and, before running out of gas, landed on a beach in Africa. A "duck" from Roberts Field, Liberia, brought gas and removed cargo to lighten the plane. Under the direction of the engineers aboard, native workers hauled ocean water to harden the sand to make a runway and Wold took off and landed at Roberts, a field operated by Pan American airways."[20]

The majesty of the British empire was built on knowledge of and respect for the sea. Even before World War II, regular over-water meteorological flights were established, covering the open ocean around the British Isles, as far south as the Azores, west to 37° longitude and above the Arctic Circle to 70° north latitude. At the peak of recon during World War II, ten Royal Air Force (RAF) "sorties" were flown daily. *Magnum* originated in Reykjavik, Iceland; *Recipe* flew from Noupe Head, England to above the Arctic Circle, while *Bismuth* and *Mercer* originated at Tiree, Scotland. From Trevose Head, England, *Allah, Epicure "A"* and *Sharon* operated as a shuttle to the Azores; *Epicure "B"* originated from Brawdy, England, and *Rhombus* from Docking, England, to Skitten, Scotland. *Nocturnal* flew in and out of Gibraltar. Blenheim aircraft were used for most of the flights because, as one pilot explained, "If the cloud base is ten feet or more, a Blenheim can land because it is only nine feet high."

Col. Elliott Roosevelt, with the Eighth Air Force, directed tactical and synoptic recon flights out of England in 1943-44. Roosevelt's squadrons flew B-24 and B-17 aircraft, and teamed on flights with the British.[21] The RAF, with more experience in weather reconnaissance, helped train American Eighth Air Force crews after their arrival in England by providing observers to fly with American crews.[22]

The B-17 had proven itself in synoptic and tactical recon in England, and as weather reconnaissance squadrons were being formed, the B-17 was the most often requested aircraft for recon. Depending on the theater, weather B-17s were stripped of gun turrets and armor, and fitted with bomb bay tanks which allowed 3600 gallons of fuel on board. The Boeing B-17 was the famed *"Flying Fortress,"* the old war bird that earned battle honors throughout Europe and the world, the first bomber to attack the Japanese in the Philippines and the bomber of choice of the Eighth Air Force against Germany. The plane had a wing span of 103 ft., 9 inches, was over 74 feet long and about 19 feet high, with four 9-cylinder 1200 HP Wright Cyclone engines. Important to weather recon was its ceiling of 35,000 feet, the 2,000-mile range and speed of nearly 300 MPH.

It was unexpected that weather crews would have fatal missions. However, in the late evening of 22 September 1944, a B-24 returning from an *Epicure* mission crashed while landing at Watton, England, killing the 18th Weather Squadron's 2nd. Lt. John H. Mackling, weather observer, and M/Sgt. Richard W. Stoodley, forecaster.[23] On 17 March 1945, a B-17 of the 652rd Bomb Squadron was forced to ditch off St. Evil, Land's End, England.

Interesting sidelights to any war have been friendly exchanges between enemies. For example, on recon flights out of St. Mawgan, British Lancasters often encountered German flights from Brest. As the planes met, the pilots would salute each other by dipping wing tips—a flying courtesy even today. RAF squadrons were ordered not to attack German weather aircraft because the British could decipher the German code, eliminating the need for additional RAF weather flights.[24] Sending weather observations in the clear prevented many planes from being shot down because both sides could use the information.

Weather reconnaissance was slowly proving its value in combat. Success in military strategy often comes with new ideas and by breaking long-standing tradition. Hard-bitten generals found it difficult to accept the fact that aircraft flying through clouds might contribute to war tactics. Fighters had never been considered "cloud watchers," or "weather birds," until the "Orphans of the Storm" Mercury flights of the Ninth Weather Reconnaissance Squadron, changed these ideas. Gen. Tooey Spaatz credited the Orphans with saving the Air Force $30 million in aircraft fuel costs for June, July and August 1944. In December 1943, Maj. Gen. Samuel Anderson could find no use for reconnaissance missions. Seven months later, after D-Day, he refused to send out a single bomber without first getting weather reports from recon planes.[25]

Since the war silenced ships at sea, and weather information from enemy-occupied areas was cut off, most of the normal world-wide weather observation posts were eliminated, denying meteorologists critical information for forecasts. The new war was in the air—battlefields had become three dimensional, and air commanders were vitally interested in the weather their aircraft would have to fly through to accomplish their missions.

The atmosphere is the most treacherous of all terrain. "If a ground commander was confronted with mountains changing height, rivers altering course, ground moving up and down, his terrain problems would be like those of the air commander."[26] To learn the changing conditions in the atmospheric battle grounds, the most obvious sources of information would be from specially trained air crews assigned to check weather over ferrying routes, bombing targets or gathering weather data from areas where storms and critical weather were being formed. To the air commander, weather hundreds of miles away could be of much greater significance than weather over his headquarters.[27] This problem led to organized reconnaissance flights over oceanic and Arctic areas where there was no weather reporting to provide data for synoptic charts of the weather forecasters.

Weather recon had a military and civil precedent—it wasn't a new idea. During and following World War I, observation balloons and aircraft were used to gather upper-air information, and weather instruments were attached to the outside of commercial aircraft. Civil airlines, in cooperation with the Weather Bureau, set up weather stations along their routes and relied on their own pilot reports for weather information. British reconnaissance dates back to 1916, when Capt. C. K. M. Douglas of the Royal Flying Corps, 15th Squadron in France, directed his pilots to record temperature and cloud observations on every flight, and, in 1918, as part of Meteor Flight, in France, he took upper-air data for the Royal Engineers.[28] Following W.W.I, in England, a "meteorological flight" was established at the Air Pilotage School at Andover, using Bristol F2B fighters, and Gloster *Gauntlets*. In the thirties, British pilots maintained two recon flights per day, and, in the period from December 1934 to December 1935, pilots "achieved one full year of operation without missing a climb." These were called THUM (temperature and humidity sounding) flights.[29] The first U. S. Navy weather reconnaissance flight was made over the Gulf of Mexico, on 4 May 1918, by Lt. W. F. Reed, from the Naval Air Station at Pensacola, Florida. By June, Lieutenant Reed was making regular flights to provide upper wind velocity and direction and reports were sent to weather stations for forecasting.[30]

With the world's attention focused on Europe, war in other parts of the world attracted little attention. However, the 17th Weather Squadron in the South Pacific, was fighting a bitter battle with the Japanese. The 17th was activated on 18 September 1942, at McClellan Field, California, and sent to Auckland, New Zealand, arriving there on 22 November. The squadron was responsible for weather forecasting from New Zealand to the equator and from Australia to the Cook Islands. To supplement their forecasts, it was necessary to have reconnaissance data, so weather observers flew on bombing missions—they called themselves *"weather merchants."* The first regular reconnaissance in the South Pacific began in April 1943, from Henderson Field, Guadalcanal, and flights were made in Navy SBD's, Catalina PBYs and Venturi PV1s—Navy aircraft with Army observers. The weather observers logged thousands of hours over Japanese-controlled territory making weather observations for bomb target forecasts. Daily pre-dawn weather flights were made in a torpedo bomber from Green Island, to Rabaul to Simpson Harbor. From these flights, on-the-spot weather conditions were sent back to Bougainville just before the daily strike-bombers took off. In 1944, Army weather observers began flying on northern search areas in Navy PB4Y planes around Truk, a major Japanese base. Weather information was transmitted in Navajo Indian language code.[31] These were called the Navajo Code Talkers.

Weather flights were not nicely-nice missions. Lt. James Van Dyne, Akron, Ohio, who logged seventy missions and five hundred hours of weather reconnaissance over enemy-held territory, described one weather search mission when his crew discovered Japanese unloading ships. They made a pass and dropped bombs. Van Dyne writes, "Four Japanese Zeros, in formation, rose up from our left with a flight path parallel to ours. I vividly remember the gold mottled paint and the big red ball. . . . I could distinctly see the pilots looking us over. For fifteen minutes—it seemed like an hour—they made frontal and rear passes. I was firing from the left waist position, . . . we escaped with minor damage, no one was injured and we did not hit any of the enemy."[32]

L. D. Clark, 17th weather observer from Tucson, Arizona, flew night B-24 search missions out of Munda, New Georgia. For luck, just before takeoff, his crew would gather at the nose of the plane and, at the top of their voices, sing an old ditty:

> Oh, I was strolling in the park,
> And goosing statues in the dark.
> If Sherman's horse can take it
> Why can't you?

Consolidated B-24 bombers were used for weather recon by the Eight Air Force in England, the 17th Weather Squadron in the South Pacific, and by Col. Nicholas Chavasse's 655th Reconnaissance Squadron flying weather for the B-29 raids over Japan. *Photograph courtesy of Howard E. Lysaker*

Once in the air, they were serious and, on one mission, by radar, found a huge naval force. Clark's description: "We decided not to take on the entire fleet and just work the edges. We made a few passes with no hits, and the Japanese ships sent up a fascinating array of fireworks. I remember standing behind a strip of armor, keeping all of me under protection except my fool head—I didn't get scared until we left the ships. We had taken some flack, a few bullets punctured our hide, and we limped home. We got credit for discovering a huge Japanese supply fleet and our bombers went after them the next morning."[33]

On the night of 20 March 1944, Lt. Raymond Pope, a weather officer, was killed when his crew crashed into the sea short of an airstrip at Munda. Pope's death was a tragic personal blow to the weather men. He had traded a flight with another crewman. Robert Lashbrook, weather observer from Ojai, California, wrote about Lt. Pope's death, "Together we had somehow survived a forty-three-day voyage overseas in an old banana boat; together we had volunteered for Munda, and we had shared the same tent." Taking weather flights continued voluntarily after Pope's death, but the practice of trading flights stopped. "Once the schedule was made for future flights, your fate was sealed. To trade a flight with someone else was just tempting fate too much."[34] Weather officers flew without flight pay.

To understand military weather missions, whether flown over an unseen anti-aircraft barrage, through the jaws of a bone-crushing tropical cyclone, or in the smooth quiet loneliness of white fog and clouds, you must try to feel the dread that rides on every flight. Bomber crews experienced fear and terror in the air—that was their job. But to the 17th weather men, who took missions voluntarily, nothing chews on the soul like the fear that hovers on a flight when he wasn't ordered to be there, or when he had swapped flights with a friend.

The 17th Weather Squadron, with weather observers flying on bombing missions, made a significant contribution to the war in the South Pacific. Members of the 17th received seven Purple Heart awards, three Bronze Stars, three Legion of Merit awards, forty-seven Air Medals, and a commendation from Maj. Gen. Nathan F. Twining. In addition, the 17th personnel conducted research into equatorial weather which supported the belief of forecasters that the inter-tropical front was not frontal in nature—a significant find."[35]

With the Eighth in England and the 17th already in action in the South Pacific, plans for dedicated weather reconnaissance were being made in the United States. Capt. Arthur McCartan, a pilot-weather officer, working with W/O Walter Sebode, was sent to Weather Service headquarters at Gravelly Pt., Virginia, in July 1942, to set up a Table of Organization and Equipment (TO&E) for a four-engine weather reconnaissance squadron. Captain McCartan, now living in Florissant, Missouri, explained the project, "I was to run around the new Pentagon and obtain B-17s or B-24s. Very quickly, Col. Oscar Senter, my boss, then director of the Weather Service, was besieged with calls from here and there in the Pentagon, wanting to know who this Captain McCartan was and didn't the Weather Service know there was a war on?"[36]

With everyone wanting four-engine bombers, the Weather Service "pulled in its horns" and settled for B-25Ds. In September 1942, Captain McCartan was assigned to the newly organized First Weather Recon Sq., Air Route Medium, Test No. 1, Patterson Field, Ohio. Their first aircraft was a C-45 and the second was a Lockheed Hudson, with open rear gunner's cockpit. These obsolete planes were delivered on 16 November 1942, by the Ferrying Command. There were no instructors and pilots were expected to learn to fly them on their own. No one in the unit had flown the Lockheed Hudson, an old British version with a handbrake. It was not the easiest aircraft to fly, but good enough to take four or five weather officers through fronts, icing, turbulence, rain, etc.

The crews started training themselves for weather recon by importing specialists such as Dr. Jon Shereshevevsky, a French weather expert, described as a man with enthusiasm and a helpful attitude. Captain McCartan was working under Lt. Col. Curtis LeMay of the 34th Bomb Group, Pendleton, Oregon. (LeMay later earned the name "Old Iron Pants" when he led the WW II B-29 raids against Japan.) McCartan wrote, "During the winter and spring [1942-43] we learned a lot about icing, snow-static, chasing legs of ancient radio aids, etc. But we had so many training accidents, many due to weather, that our group 'stood down' for sixty days, from 0400 to 1200, while I gave weather lectures and Colonel LeMay gave operational instructions to air crews."[37]

As these crews were training to become the first dedicated, sole-purpose weather reconnaissance squadron, route recon continued in an attempt to cut down weather losses of fighter planes being flown to England. Between June 1942 and January 1943, General Tooey Spattz's Eighth Air Force flew 920 aircraft across the North Atlantic. This was *Operation Bolero*. Only thirty-eight planes were lost, instead of the ten percent anticipated.[38] One of the reasons for the success of *Bolero* was a B-17 weather reconnaissance operation under the leadership of Col. Clark Hosmer. Lt. Col. E. G. Simensen, regional control officer of the 18th Weather Region, had convinced officials to use eight B-17s to fly route weather recon along each leg of the fighter route.[39]

On one flight, Lt. James Teague, flying one of the weather B-17s, was unable to land at BW-1, on the southern tip of Greenland, because the field was closed due to weather. He headed to BW-8, five hundred miles further north inside the Arctic Circle. Low on fuel, Teague decided to crash land on the western Greenland coast. Providentially, he discovered an emergency strip near Marrack and landed. The field was later developed and named for the pilot. When the Eighth AF *Operation Bolero* was completed, the weather B-17s were returned to combat.[40]

The North Atlantic ferrying route was a pilot's frozen hell—three thousand miles of iceberg-infested ocean from Newfoundland to Greenland to Iceland to Scotland to England. Young, inexperienced army crews flew through ice storms moving down from the Arctic Circle. Crews and aircraft had never faced such a challenge. Charles C. Bates, weather officer from Green Valley, Arizona, stationed at Presque Isle, Maine, agonized as he gave weather briefings and sent "mere kids" across the Atlantic toward England. "The big problem," Bates reports, "was that the B-17, B-25 and B-26 pilots had so little flying experience before heading out over the ocean. Weather forecasts for Greenland and Iceland were at least six hours old, and we often lost fronts until they hit the Azores. One C-54 pilot got trapped in a wind shift and ended up over Terceira, Azores. The Portuguese, officially neutral, fired a few rounds of ack-ack (well off the target) and let the plane land. In addition to flight concerns, small, seemingly insignificant problems often damaged morale. For example in forecasting from Presque Isle, a deputy commander, two years out of MIT, would raise hell when maps were drawn in UCLA style. We called him 'Leaping Leo, the Lion.'"[41]

The First Weather Recon Sq., Air Route Medium, Test No. 1, after training at Truax Field, Wisconsin, was moved to Presque Isle, Maine, and designated the 30th Reconnaissance Squadron. As the 30th began North Atlantic operations in June 1943, nine new twin-engine B-25Ds had been delivered, freshly painted with green tops and white undersides loaded with weather instruments.[42] This new squadron of B-25s was the first specifically assigned to weather reconnaissance.[43] Lt. Col. Clark Hosmer, fresh from *Operation Bolero*, was commander. Hosmer was a pilot and a graduate of the famed Irving Krick weather school at the California Institute of Technology. With nine experienced crews, Colonel Hosmer divided the North Atlantic route into three flights of three crews each. Flight A flew between Goose Bay, Labrador, and BW-1, Julianehaab, Greenland, a distance of six hundred miles. Flight B was from BW-1 to Reykjavik, Iceland, 950 miles; and Flight C was another thousand miles to Stornoway, Scotland, or into England. Occasionally, crews would fly to BW-8, inside the Arctic Circle, and continue over the ice cap to Iceland. All of these flights were over open water, paralleling or inside the Arctic Circle, or over the Greenland ice cap, across the top of the world. Crews would fly Flight A, and remain over night (RON), take Flight B and another RON, and on to Scotland for the third leg. The missions would be repeated back to Goose Bay with crews following each other on a daily basis, reporting weather for ferrying crews and forecasters. During 1943, from June to December, Hosmer's squadron had no losses due to weather and won Air Medals for the commander and his crews.[44]

The B-25D, North American Mitchell, designed as a medium bomber, performed well and was extremely dependable. The plane's long range, flight stability, speed, ease of maintenance and good performance, made it an excellent aircraft for weather reconnaissance. On one mission, Capt. Doak Weston, operations officer for the 30th WRS, flew his B-25D with one engine for 250 miles on instruments, made a letdown and landed in Iceland.[45] The B-25 saw more different kinds of service than any other aircraft in the army arsenal.

During World War II, B-25s were called on for unusual missions such as Jimmy Doolittle's raid over Tokyo, when the land-based bombers took off from the deck of the aircraft carrier USS *Hornet*. B-25s flew the first Arctic operations across the North Atlantic, the first hurricane reconnaissance in the South Atlantic, weather recon over India's Bay of Bengal, plus low-level bombing and strafing throughout the world and, because of its dependability, served as a "generals' aircraft" in many theaters. Often called "The Model-T of the Air," the B-25 rattled and groaned, but it got you there and back. Powered by two Wright Cyclone R-13, 1750 HP engines, the plane had a range of ten hours (with extra bomb bay tanks), cruised at 200 MPH with a ceiling of twenty thousand feet. It had a wing span of 67 feet, 10 inches; length of 52 feet, 11 inches; and height of 15 feet, 10 inches, carried a five-man crew and was happy to operate off of two thousand-foot runways. For long-range weather recon, the planes were stripped of all armament and 960-gallon fuel tanks were installed in the bomb bay.

The B-25D's of the 30th Weather Recon Squadron were equipped with special weather instruments for synoptic observations. Among these were the ML 313/AM psychrometer for finding temperature and humidity; ML 175 aerograph to take air speed, pressure, relative humidity and free air temperature; Frieze aerograph, a back-up for the ML 175; aneroid barometer mounted in the nose, used as a cross check for other instruments; cloud-height meter to measure cloud tops and bottoms from flight altitude; and Type C 13A free air temperature gauge for use if electrical equipment failed. In addition to normal flight instruments, SCR 718A radio altimeters were installed in all aircraft; and later, an additional radio altimeter, the AN/APN-1 was installed. These radio altimeters saved many lives and missions during let down over the ocean when there was no setting for the pressure altimeter and made it possible to safely fly over the Greenland ice cap and enter hurricanes at near sea level when pressures were constantly changing. Special AN/APQ-13 cockpit radar made weather flying easier and safer—pilots could pick their way through thunderstorms to avoid heavy turbulence, and radar was used for over-the-ocean letdowns when fog-bound coastlines in Greenland and Iceland prevented normal letdowns. Standard equipment on most multi-engined aircraft was the driftmeter to take double drift wind readings for navigation.

As the success of the 30th WRS's "short-leg" recon over the North Atlantic was recognized, other recon flights were made. Stanford Kahn, from Wilmette, Illinois, describes *Stork* flights—three Air Transport Command (ATC) C-54s flying the long trip nonstop across the Atlantic and return. *Stork* was a regular military cargo-passenger transport run, with weather observers from the 8th Weather Squadron on board. Kahn, as a weather observer, flew seventeen missions on these flights, from late 1943 to 1945.

The route was normally from LaGuardia Field, New York, to Stephenville, Newfoundland, to Prestwick, Scotland, with the west-bound flight stopping at Keflavik, Iceland. The ATC planes were flown by American Airlines (AA) or Trans World Airlines (TWA) flight crews.

Charles Bates was active as a weather forecaster in England, and flew as a weather observer on many *Stork* flights. Bates tells about the very first experimental flight from Stephenville to Prestwick. Capt. Jack Catchings, senior AA pilot, was flight commander on the ATC C-54, No. 281. The date was 2, 3 December 1943. Through the years, Bates has carefully kept a copy of his route forecast and flight log for this first C-54 weather recon flight. The flight log records a long, labored, bad-weather take-off with the C-54 heavily overloaded with P-51 drop tanks for raids over Berlin. Just clearing the end of the runway, Catchings turned to Bates, and muttered, "It's things like that that make your ass-hole twitch!" Other crew members signing the log were Capt. Percy E. Cunningham, AA pilot, Art J. Vasold, E. T. Fisher, and Marvyn Carlton, navigator, all AA crewmen.[46]

On *Stork* flights, the eastbound distance from Stephenville to Prestwick was 2,024 miles with a flight time of 11 hours and 50 minutes at 170.9 knots per hour. Westbound, the time was 14 hours and 31 minutes at 139 knots—both flights at nine thousand feet. The only unusual entry in the typewritten log described the take off from Newfoundland. In an appended handwritten note, Captain Catchings wrote, ". . . thick snow during take-off; . . . air speed lowered to below 120 MPH and altitude dropped from 200' to 100' . . . caused by snow sticking to the wings." Under this notation was, "Came relatively close to crashing!"[47]

Weather reconnaissance pilots spent their military careers flying severe weather. That that was their job and they were good at it. It was a spooky assignment—piloting an aircraft in actual weather, on instruments, with tons of machine, humanity and fuel, hurtling at more than 200 MPH through a visual nothingness—instrument flying was the most threatening part of any flight. Pilots, at the beginning of the war, received very little instrument instruction in actual weather situations. Students were trained on the Link Trainer[48] and were given non-threatening flight instruction with the student pilot under a hood while the instructor kept careful watch. For example, I was an instructor pilot in B-25s at Columbia Army Air Base (CAAB), Columbia, South Carolina, and our students were given no flight instruction in actual weather. In 1944, with over two years experience as an Army pilot, I had a puny minimum of weather experience and was extremely uncomfortable when I could not see the ground.

However, actual experience has always been the best teacher. On 5 May 1944, orders came directing me to deliver a B-25 D to Grenier Field, New

Hampshire, and fly by Army transport to Presque Isle Army Air Base, Maine, for thirty-days temporary duty (TDY)[49] with the 30th WRS. I was to fly the North Atlantic as copilot to get weather experience. This directive was "By order of Lt. Col. Clark Hosmer." I boarded an Army C-46 contract carrier, with a Pan American civilian crew for the flight to Presque Isle. The weather was near freezing with low clouds, light snow and rain. I assumed we would certainly wait for better weather. But, no! The civilian crew came on board and, to my utter disbelief, taxied out to the take-off runway—this would be my first experience with real weather and I was sure we would all die. After a wet, soggy takeoff, with moderate bumpiness, we reached cruise altitude and were still in the clouds—rain was streaking across the cabin windows with little ice crystals forming along the edge of the Plexiglas. I looked around the huge plane, and scrawled in chalk on the aluminum bulkhead of the pilot's compartment was "Kilroy was here!"[50]

After reaching cruise altitude, flying in bumpy clouds, I went forward to visit the crew. As I entered the cabin, the pilot was asleep, the plane was on autopilot, the copilot was reading *Time* magazine, and the flight engineer was watching the instruments and looking bored. This was one of those experiences in life when you grow up instantly! Here at fifteen thousand feet, flying in a limbo of black clouds, with rain and ice in the worst weather I had ever seen, the crew was half asleep and the plane was flying itself. I stood in awe—a mix of anxiety and respect—the coolest situation I had ever seen—pure Zen. It is important to mention this incident, because regardless of training and experience, this is how weather pilots are made. From that instant, any feeling that weather flying was a before-death experience was forever erased. In that one second of time, without having touched the aircraft controls in bad weather, I became a weather pilot! Fear of weather was gone! There was respect, yes! But fear? No! Later, on many occasions, I learned that the weather that day was not the worst I would experience along the North Atlantic and in dozens of other places throughout the world. This is the repeated story of parallel careers of hundreds of weather pilots who found experiences that no one dreamed could exist and discovered flight situations beyond rational belief. We were alone and often afraid in a world we could never imagine. We were out to prove, as men and women have always tried to prove, that there is no limit to the frontiers of flight or to human daring.

The story of the 30th WRS's flying in the North Atlantic is best told through personal experience with the squadron. At Presque Isle, I was assigned as copilot to Capt. Doak A. Weston, the wildest, most uninhibited, most competent pilot I had ever flown with, who, by coincidence, was a former fellow instructor, in 1943, at Columbia AAB. At Columbia,

Weston had witnessed the death of his crew the year before on a B-25 skip-bombing training mission over Lake Murray, South Carolina. His student pilot dug a wing into the lake and all on board except Weston were killed. He left instructing to join weather reconnaissance. My TDY with the 30th and instruction by Weston was short—just thirty days—time enough for two round trips from Goose Bay, Labrador to BW-8, to Iceland and Stornoway, Scotland. The world had become a series of strange names and far away places—there was wild and primitive romance in names like Goose Bay, Gander, Narsarssuak, and Bluie West.

Our first flight was from Goose Bay, on Bluebird Flight, to BW-8, above the Arctic Circle in Greenland. Now BW-8—Bluie West-8—is a very special place in the world, named by Arthur Merewether, early weather pioneer, who thought frozen Greenland just looked "bluie."[51] BW-8 is a small ice-locked air field with steel mat runways at the end of Søndre Strømfjord, inside the Arctic Circle. As we approached the jagged coast line of Greenland, we were encountering flight situations that would be absolutely insane in civil aviation but routine in weather flying. It was only by piloting skill, luck and faith that planes landed at BW-8.[52] The approach to Greenland from the west was a breathtaking scene. After the letdown over the North Atlantic, the huge mass of Greenland grows darker and darker through the mists and clouds. There lies the largest cake of ice in the world, blanketing a frozen continent one fourth as large as the United States, three-fourths inside the Arctic Circle, where the sun shines only four months of the year. To get to BW-8, we had to find a hidden fjord.

As Captain Weston made the descent midway up Greenland's west coast, and fixed his position over Semitak Island, he had a life-and-death decision to make. Two miles from Semitak are the mouths of three fjords and only one leads to BW-8. I had been told that once an aircraft enters any of the fjords it is impossible to turn around. But in my military flying, when death was the alternative, I have seen many impossibilities that quickly become easy. Thirty miles into the correct fjord is a wrecked freighter—the only positive landmark. With trust and ignorance as my only comforts, I calmly accepted Weston's decision that he had found the right fjord, but became more and more concerned as I watched the sheer glacier-cut walls of the fjord move closer and eventually so close that a turnaround was impossible. Then, we saw it! Never has the rusting hulk of a wrecked freighter been so beautiful! The clouds dropped close to the water, and we were flying in a tunnel of rock, water and clouds, fifty feet above the icy black fjord with floating ice chunks as large as a small house.

At the end of a left turn in the fjord, BW-8, a tiny, scruffy, weather-beaten landing field suddenly appeared—more like an Eskimo village than a military installation. Pitiful as it was, BW-8 looked as majestic as New York's Great

Flight across the great Greenland ice cap was an alternate leg of weather recon B-25s flying the North Atlantic. Craggy mountain peaks pushed through the snow and pilots fought blinding haze. Several aircraft have been lost over the ice cap. *Photo courtesy of W. A. Merrill*

White Way. The landing strip was a long steel mat; wheels went down, and we hit the mat with a shuddering crash for a normal landing. We spent the "night"—although it was daylight around the clock—with the pale sun describing a circle around the horizon. For breakfast, there were eggs (some weather crewmen said they were duck eggs) that we were told had been frozen in an ice cave for a hundred years. Later, airmen who had been stationed at BW-8 were disappointed that stateside fresh eggs "had no taste."

Taking off the next morning, over the Greenland ice cap, the sight was unbelievable. The huge mass of ice, the last primitive remains of the most recent ice age, is about ten thousand feet thick, one thousand miles long, and covers Greenland except along the coasts. Small mountain peaks poke through the crust of ice, and snow drifts pile high like sand dunes in the desert. Pilots, even on sunny days, were forced to fly instruments because there was no horizon, as ice and sky blended together in a dazzling never-never world with no visual references. We approached the ice cap with apprehension. On the operations bulletin board, at Presque Isle, was this story we had all read:

In early 1942, Col. Norman Vaughan, director of Search and Rescue in the North Atlantic, reported, ". . . a plane [C-47] flew into a fall
of fresh feathery snow that stopped it so gently that for several minutes, the crew thought they were still flying, although the airspeed
needle had quietly swung to zero. They were so firmly nestled in
snow the steady-cruising propellers had blasted ten-foot furrows for
half a mile and the fuselage was buried to the windows."[53] In November 1942, a B-17E, searching for the lost transport plane, crashed
near a research station set up on the ice. First Lt. Max Demorest, a
forecaster from Flint, Michigan, and S/Sgt. Don E. Yetley, set out
from their station on motor sleds to locate the bomber. Then tragedy
struck. After they found the B-17, Lieutenant Demorest, bringing up
the sled, broke through a snow bridge and fell more than one hundred feet into a deep crevasse and was never found.[54]

Flying at eleven thousand feet, after we had leveled off over the ice cap, the
usual jovial Weston was deadly serious. He kept his eyes on the instruments, especially the radio altimeter which showed actual height above
the mass of ice. Only when we could see water below, and knew that we
were clear of the ice cap, did Weston relax.[55] Flying over this great mass of
blinding ice and across the North Atlantic ferrying routes were the most
maturing experiences any pilot could have. Pilots learn to respect all weather,
and we learned that almost any weather can be safely flown with the aircraft
and trained crews of World War II. Otherwise, weather reconnaissance would
have been impossible.

As I write this history, fifty-two years after flying over the Greenland
ice cap, I am amazed that many of the things found in World War II were
even then predicting changes in the earth's weather. We poked our planes
into the vast, unknown weather breeding places and never understood
what we found. Today, with the blizzards and floods of 1996, the fourteen
hurricanes of 1995, the floods, winds, and weather violence, it is evident
that something was and is happening with the world's weather. For example, consider this story: In August 1928, a small plane took off from
Rockford, Illinois, to fly nonstop to Stockholm, Sweden. After missing his
refueling station over Greenland, the pilot landed on the great ice cap—he
and his copilot were able to walk to safety.

Sixteen years later, in 1944, a Navy reconnaissance plane was flying
over the ice cap and found the 1928 plane, nosed over by the wind. In the
long years the plane had stood in the frozen waste, it should have been
buried deeply in the snow. The evidence was clear—the howling blizzards
were no longer building up the Greenland ice cap. Is our climate changing? Evidence indicates that the world is getting warmer, and weather
cycles are showing strange variations. Why? Scientists understand day-to-

day weather. They know why hurricanes roar in from the sea, why tornadoes spin violently across the land, why blizzards rage and heat spells come
and go. But they do not know what forces of nature create sweeping long
range climatic upheaval.[56] Weather reconnaissance crews discovered unknown weather phenomena, but they never really discovered the reasons
or the long range implications of changes in weather.

In addition to learning to fly in Arctic weather, crews made ingenious
engineering discoveries in the field that made it possible to operate aircraft in the sub-sub-zero conditions. For example, when we landed at BW-
8, before cutting the engines, Captain Weston held a toggle switch "on"
until the engine oil pressure dropped to zero and then cut the engines. This
was a technique called "oil dilution." In the extreme Arctic cold, a plane's
engine oil would congeal in a few hours, and it would be impossible to
turn the engine over for the next start without pre-heating. Oil dilution,
while the engine was still hot, poured gasoline into the crankcase until
the oil was thinned to almost water consistency. Next morning, the engine would turn over easily and start with no problem. During the engine
pre-flight warm up, the gasoline would vaporize and the oil would regain
its viscosity.

Oil dilution was discovered by an unknown flight engineer and became
standard procedure for extreme cold weather flights on many aircraft. Before oil dilution was discovered, two hours of preheating with a portable
gasoline heater were necessary before the engines could be turned over.[57]
On fields without preheating facilities, after shutting down the engines,
and, while it was still hot, the oil was drained out on the snow where it
would congeal. It was "rolled up," placed in a drum ready for heating the
next morning, to be poured back into the engine when it became liquid.[58]

Personal encounters with an invisible enemy along the North Atlantic
illustrate the mystery and intrigue of warfare. German submarines in-
fested the icy water, and often attempts were made to sabotage aircraft by
sending false radio signals to lure crews to their deaths over the open water. Radio operators, called "sparkies," could recognize personal radio
techniques—the radio operator's "fist"—as they tapped the signal keys.
James Johnson, Reading, Massachusetts, remembers one mysterious incident along the North Atlantic: "While working my ground station to an
aircraft, I was getting approximately R3-S3 (R is readability and S is signal
strength, with R5-S5 the best). Suddenly a signal came through as if the
sender was in my lap. I had never worked that 'fist' before! A strange aircraft was calling in! I alerted base weather, and went about my business as
fighters screamed off the base. But I never knew what happened."[59]

BW-1, on the southmost tip of Greenland, was also difficult to fly into.

Pilots coming in from the Narsarssuak Fjord, over Julianehåb, had to land slightly uphill on the iron-matted runway. A hot shot pilot, flying a Mosquito bomber to England, decided to come in from the ice cap and land downgrade. The landing ended in the fjord and the pilot was rescued in time to watch his plane sink in the water.[60] At times when BW-1, Greenland, or Reykjavik, Iceland, were below let down minimums, pilots would fly to the field beacon, turn away from the coast and letdown over the ocean using the radio altimeter, until they could see the water. Then, skimming just under the soupy weather, the pilot would fly back to Julianehåb or Reykjavik by DF signal, cockpit radar and the radio altimeter. After a long mission, flying away from your landing field and letting down over the ocean through icy thick clouds is an act of faith. In a fifty-foot-per-minute letdown, as the clouds begin to break, everyone strains to see the water, and you desperately want to dive through the holes to get below, but experienced pilots know that they fly instruments until completely clear of the clouds, and then fly contact back to the base."[61]

Aircraft weather reconnaissance continued to gain respect. One significant incident, in 1944, involving a C-54, proved that weather reconnaissance was earning trust. Glenn Bradbury, Fort Worth, Texas, explains, "In September 1944, while on duty as forecaster at the St. Mawgan, Cornwall, ATC station, England, an Air Force major requested clearance directly to Washington, D.C. I had to tell him that, because of the little weather information we had, we could only clear him as far as the Azores or Iceland. The major, who was the pilot, asked to see the weather maps.

"He reviewed the isobars around a large low-pressure system, 1,220 miles to the west, north of the Azores, and was extremely interested in reconnaissance aircraft reports on the map which clearly identified the location and movement of the storm area. He insisted that he be cleared. Being a second lieutenant, I called Captain Sergeant, who also refused to sign the clearance. The major asked if we would permit him to sign his own clearance. We yielded and supplied all of the information we had."[62]

"The major took off in the C-54 at 2000 hours, destined for 'Newfoundland and beyond.' Three days later, we read in *The Stars and Stripes* that a Major Myers had flown *The Sacred Cow*, President Roosevelt's personal plane, with Secretary of War Henry L. Stimson on board, non-stop from England to Washington, D.C. Major Meyers had flown a great circle course on the north side of a low pressure system taking advantage of tail winds across the Atlantic for the first official non-stop flight from England to Washington, D.C. Had it not been for those weather reconnaissance reports plotted on that Sunday evening map in St. Mawgan, it is unlikely that *The Sacred Cow* pilot would have signed his own clearance for the non-stop flight to D.C."[63] This flight opened the way for "pressure pattern" flying, taking advantage of tail winds by following the circulation around low and high pressure weather systems for long-distance trans-oceanic flight.[64]

The winter of 1942-43 was too severe for ferrying traffic across the North Atlantic. The accident rate rose from 2.9 percent in September to 5.8 percent in October, and ATC suspended winter operations and directed traffic to the South Atlantic.[65] The South Atlantic route was nearly twice the distance, but much safer. The 30th WRS stationed recon crews in Trinidad, Belem and Natal and one flight on the Azores. Col. Clark Hosmer, commander of the North Atlantic 30th, established weather recon flights along the top of South America, to Ascension Island and to Liberia, Africa.

Except for hurricane reconnaissance which began in 1944, weather reconnaissance along the South Atlantic was not a high priority. The Air Force depended on Panair de Brazil, a subsidiary of Pan American Airways, for weather information.[66] The South Atlantic "aerial highway"—Miami-Natal-Ascension-Africa-India—was blessed with excellent flying weather almost the year around, except during hurricane season, when the winds turned easterly and the storms developed in the doldrums.

In July 1943, the 9th Weather Squadron was reorganized and became the 2nd Weather Squadron with headquarters at Natal, Brazil, under the command of Lt. Col. James Baker. Forecasters flew on regular ATC flights along the route, making in-flight weather reports. This procedure was not successful since aircraft observers in flight were sending reports on weather they were then experiencing—too late to provide information for forecasting. However, when area synoptic reconnaissance began in 1944, the reports were so successful that weather became negligible as a cause for aircraft losses. Weather stations at Waller Field, Trinidad; Belem, Brazil; and Morrison, Florida, each had two aircraft for area reconnaissance.[67]

Life along the top of South America was leisurely with plenty of booze and other refinements of life. In fact, South American bars were the only ones in the world where airmen constantly requested more Coke in their rum drinks since rum was cheaper and more plentiful than Coke. Some of the world's best rum was not imported into the United States, and air crews made regular "rum runs" from the South Atlantic. Kent Zimmerman, weather pilot from San Antonio, Texas, reports that 150 proof rum was a favorite of crews.[68] Canadian Club blended whiskey, in the Bahamas, was selling for $28 a case.

Colonel Hosmer, weather pioneer, now fifty years later, confesses that he made the greatest mistake of his life in Brazil. A Weather Wing inspector sent him a copy of an inspection report on South Atlantic weather recon. The report included the comment: "... having completed their early morning flight, Hosmer's crew reported to the weather officer and 'retired to their Brazilian boudoir.'" Hosmer didn't like the possible interpretation of that phrase. Instead of asking the inspector for clarification or telling Col. Oscar Senter, Weather Wing commander, he thought the phrase was unjustified, he and Capt. Doak Weston, his operations officer, "got carried away in drafting a letter to Colonel Senter, upholding the integrity of his crews and his belief that the phrase was out of order." Weston almost cried with regret that they did not call the report "snide." In addition, Hosmer compounded his "blunder" by indicating that he had sent a copy to all regional control officers. Later he learned that Colonel Senter's deputy had never seen the commander as mad as when Hosmer's letter embarrassed him in the eyes of his subordinates. "This forever ruled out further serving in the Weather Wing and any chance of promotion," Hosmer wrote. However, apparently Colonel Senter ignored the incident as Hosmer continued to receive excellent ratings.[69]

Col. James Baker formed the Second WRS in 1944, which was sent to the China-Burma-India (CBI) theater and assigned to the Tenth Weather Squadron. Baker's replacement in the South Atlantic was Maj. Arthur McCartan who had organized the 30th across the North Atlantic.[70]

2

Developing a Tropical Cyclone Warning Service
1943-1947

*H*unraken was the "god of stormy weather" to Guatemalan Indians; Carib Indians of South America called it *huracán,* the "big wind," and to the West Indies Taino Indians, *huracán* meant "evil spirit." From these ancient words, meaning "all that is big, stormy and evil," we get our word hurricane.[1] A Caribbean native, in 1944, fleeing with his family to the hills north of Aguadilla, Puerto Rico, told stories he had heard from his parents and his grandparents of death and destruction from the lashing winds of *el huracàn muerte,* the "the storm of death."

Thomas Morton, writing in the *New England Canaan,* reported the first recorded hurricane in America was on 15 August 1635 and ravaged the Plymouth Colony. No details were given.[2] Hurricanes and typhoons are the most destructive weather phenomenon known. "Tropical cyclone" is the generic name for a non-frontal, low pressure system, hundreds of miles across, which develops over tropical water with a counter-clockwise circulation and winds from 64 to 129 knots. West of 180° longitude, tropical cyclones are called *typhoons.* In the Indian Ocean they are called *mauritius hurricanes* or *cyclones;* the "big winds" are called *bagyuio* in the Philippines, *reppu* in Japan, *willy-willy* north of Australia and a*sifa-t* in Arabia.[3]

Weather writers and historians grasp for superlatives when they try to describe this most awesome of all the earth's storms. One writer compared the atom bomb to a hurricane—"like a flea to an elephant." Dr. W. J. Humphreys, veteran of the United States Weather Bureau, said, "A full-fledged hurricane generates more energy than one thousand atomic bombs. . . ." President William McKinley: "I am more afraid of a West Indian hurricane than I am of the entire Spanish Navy."[4] We of the present often think of the hurricane as a modern scientific phenomenon, complete with the drama of radio, television, and satellite warnings, telling us where the storm will hit and how strong the winds will be—all in good time for preparation or evacuation. Hurricanes have been a threat for centuries, and eons before that, when the storms lashed ashore on undeveloped and sparsely inhabited coastal areas. Now that there are more of us, with our billions in property along the coasts, the storms are a greater threat.

On 8 September 1900, at Galveston, Texas, a huge hurricane moved ashore from the Gulf of Mexico with its impending presence as the only warning. The hurricane completely covered and destroyed the sea-level island in just a few minutes, took six thousand lives, and became the greatest natural disaster of this century. Forty-three years later, in the early morning of 27 July 1943, another Gulf hurricane aimed its winds and rain toward Galveston and Houston. People, in a frenzy of fear, clogged highways into the interior of Texas. At Bryan, Texas, 150 miles north of Galveston, at a small airport, Col. Joseph B. Duckworth, to satisfy his "personal curiosity," decided to make an experimental flight into the hurricane, something no one had dared before. Colonel Duckworth was director of the Army Air Forces Instructors' School at Bryan, where Army pilots learned instrument weather flying. Duckworth had often wondered if a pilot and aircraft could withstand the rigors of a hurricane,[5] and he bet his life and aircraft that it could be done.

Between 2 and 7 p.m., flying a single-engine, AT-6 *Texan* trainer, Colonel Duckworth made two flights around and through the center of the hurricane.[6] Lt. Ralph M. O'Hare, navigator, was on the first flight and Lt. William H. Jones-Burdick, a pilot weather officer, was Duckworth's passenger on the second flight. Aside from a rough ride, there were no unusual flight problems on either trip, and Colonel Duckworth proved that an aircraft and a skilled pilot could safely fly through a hurricane! Before these two historic flights, the Army had refused to send air crews into a hurricane, considering that it would be a sentence of death.

Colonel Duckworth described his experience: "Upon entering the overcast at the edge of the storm . . . [after] a series of local squalls, the air was smooth, except . . . ordinary roughness associated with such squalls." In the storm, the air smoothed out completely and rain became steady and heavy. As the storm neared Galveston, rain increased and rain static was exceptionally heavy. However, radio transmission was good and the local radio station warned Duckworth that "Galveston was experiencing a hurricane."

Turning north from Galveston, where Duckworth thought the "eye" of the storm might be, he encountered rough flying—choppy, bumpy turbulence and a few updrafts—but still far less severe than a thunderstorm. Duckworth wrote, "As we broke into the eye of the storm, we were contact,[7] and could see the sun and the ground. The eye was like a leaning cone and observation of the ground showed considerable wind. We made no attempt to come close to the ground due to the possibility of exceptional turbulence from ground friction."

On the second trip, Colonel Duckworth and his weather officer pilot encountered the same conditions. A deeper penetration was made to check the temperature in the southeast quadrant where the free-air temperature went from 8°C. to 23°C. They flew through the eye again and saw that the storm was affecting Houston more severely—the radio station had been forced off the air. Duckworth's summary was that his flights through the hurricane were no more uncomfortable, the rain was no heavier, and there were no more severe updrafts or choppy air than in a good, rough thunderstorm— no worse than an unstable warm front. Duckworth's final statement was, "The only embarrassing episode which could have occurred would have been engine failure which, with the strong ground winds, would probably have prevented a safe landing, and certainly would have made descent via parachute highly inconvenient."[8] Not realizing the great discovery he had made, Colonel Duckworth apologized that his mission was not planned to satisfy a larger objective than personal curiosity.[9] However, Duckworth's "flight of curiosity," in 1943, was considered the first "official" hurricane reconnaissance flight.[10]

The first "honorary" weather recon aircraft, with only two missions, was the North American AT-6, "Texan." The "AT-6" (the Navy "SNJ" and the Canadian "Harvard,") was the military's best training plane and one that most Air Force pilots wrestled with in advanced flight school. It was a single-engine, low-winged monoplane, as tough as a Texas mustang, and as air worthy as a hawk—a small plane, 29 feet long, 11 ft., 8 inches high with 42-foot wings,. The one engine was a 9-cylinder, 600 HP Pratt & Whitney that carried the plane to 20,000 feet at 180 MPH, with a range of about 750 miles. The Texan was equipped with all-flight instruments, normal and high-frequency radios. In 1942, an AT-6 cost $25,672.[11] With thousands of other pilot trainees, I flew the AT-6 for many hours in training, instrument instruction and weather flight. While student pilots blasted away, I nervously towed targets in the only plane which I felt I was "putting on" rather than getting into. I know why Lindbergh and his plane were "we."

Duckworth never doubted the ability of the AT-6 to buck the hurricane— if there had been any doubt, it would have been his ability to control the aircraft in the storm and weather the unknown perils of tropical cyclone flight. The AT-6 still flies today and, when available, enthusiasts pay up to $200 an hour to ride in this 55-year old antique of the air, the first aircraft to officially fly through a hurricane.

Hurricane *Andrew*, in 1992, has become the storm by which others are measured. A hurricane reconnaissance WC-130 from the 815th Weather Flight at Keesler AFB, Mississippi, flew into the storm more than forty times. *Photograph courtesy of John Pavone*

The two greatest fears from tropical hurricanes are loss of life and property damage, including crops. By contrast to the 1900 Galveston storm, the most damaging hurricane of this century, to date, hit the tip of Florida on 24 August 1992. West Indian Hurricane *Andrew*, roaring in from the South Atlantic, wreaked havoc throughout Southern Florida and caused from $20-$25 billion in damages[12] with some reports of damage over $30 billion. The storm destroyed Homestead Air Force Base and crippled Miami International Airport. Two $14.3 million F-16 fighter aircraft were crushed when the huge doors of the hanger in which they were "safely" sheltered fell on them. All major buildings on the air base were destroyed.[13] While *Andrew* was more destructive then the total of all storms since 1900 and, with millions of people in its path, there were only twenty-two deaths.[14] The difference in the two storms was that, in 1900, Galveston had no warning, and, in 1992, the people of Florida knew *Andrew's* path in advance. The Weather Bureau and storm-tracking aircraft had been watching and plotting every movement of *Andrew* long before it hit Florida.

Capt. Robert Pickrell, commander of a WC-130 hurricane recon aircraft, and others of his 815th Weather Reconnaissance Flight, from Keesler AFB, Mississippi, had "lived" with *Andrew* for days as it moved toward the coast. Weather aircraft penetrated the hurricane more than forty times in seventeen missions, each time finding its precise location, direction and speed of movement. When the huge WC-130 *Hercules* flew through the wall eye, Captain Pickrell reported his instrument panel vibrated so violently that he was unable to read the instruments. Pea-sized hail stripped the paint from the leading edge of the wings, and Pickrell was worried that the engines might stop running. Lt. Col. Gale Carter, chief meteorologist for the 53rd WRS (AF Reserve), was on the flight and reported severe turbulence, lightning, grapple (soft hail), and winds at 170 knots (147 MPH). As the plane flew over the crippled air base, Carter reported seeing "blue flashes from arcing power lines."[15]

Dr. Elbert W. Friday, director of the National Weather Service, in a statement before the U.S. House of Representatives Committee on Science, Space and Technology, in 1993, said, "It is interesting to note that the major loss of lives occurred early in this century. The major economic losses occurred late in this century. The [later] storms . . . were of the same . . . and in many cases of greater severity, than those that occurred earlier in the century. [What] you see [in] the difference between the number of storms, the frequency of storms, the intensity of storms, and the loss of life, represents the evolution of a warning system in this country."[16]

In the United States, before W.W.II, hurricane warnings were uncertain and there was increasingly impatient pressure from citizens, living along the coasts, demanding more accurate advance notice of the movement of storms. Officials in Washington and the Weather Bureau were concerned. With six thousand deaths in Galveston in 1900 in their distant memory, people also remembered the great Florida hurricane of September 1928 that killed 1,836 people. And they remembered the "grandmother of all modern hurricanes" that struck New England on 21 September 1938, with a lower loss of life, but property damage exceeding $330 million—a third of a billion dollars. As the New England storm moved across the coast at fifty miles an hour, one hundred mile-per-hour winds picked four million bushels of apples; carloads of onions were sent floating down the Connecticut River, half a billion trees, including a million treasured shade trees, were destroyed or damaged, twenty-six thousand automobiles were crushed, and windows of houses were whitened by salt spray 120 miles from the sea, as far as Vermont. Wires were knocked down and utility companies used ten thousand miles of wire and 750 miles of underground cable to restore service—messages from New York to Boston were sent by way of London.[17] A way of life had been altered for all time.

Citizens complained to Washington and President Roosevelt decided to call a White House conference in August 1937. Because of the war in Europe, the President was not able to be in charge of the conference and asked his son James Roosevelt to convene the group. Before the conference there had been some thought that the Coast Guard might send cutters into hurricanes to make observations—a plan discouraged by the Weather Bureau and the Coast Guard. The Weather Bureau's position was that if there was enough information to know where to send the cutters, then they would not need the cutter's reports. The Coast Guard's position was that it was their business to keep ships out of storms—not to send them deliberately into danger.[18]

Before the conference, the President gave James Roosevelt a note which said, "Make it clear that I will veto . . . a [Coast Guard] hurricane patrol. The safest and cheapest thing would be a study of hurricanes from . . . points on land and around the Gulf of Mexico instead of sending a ship into the middle of a hurricane." It was believed at this time that coastal radar could scout approaching storms. However, before the meeting, James Roosevelt was not able to make the President's position clear to the government's delegates. At the conference, much to the surprise of everyone, a Navy admiral pledged support for the program and said he would send Coast Guard cutters into the storms. The conference came to a quick close. As a result, for the next two years, the Weather Bureau would dutifully request observations, a Coast Guard cutter would be sent out, as agreed, but as soon as the ship was near the storm, the captain would abort the mission and refuse to send his ship into danger.[19] The Weather Bureau did not complain because the storms moved faster than the cutters and observations would have been useless. At one point, in 1941, the Weather Bureau had considered hiring commercial air crews to scout hurricanes, but Congress repeatedly turned down a request for funds. Sending military aircraft into hurricanes had been a consideration for many years, although no one knew what would happen to the plane and no one wanted to venture into the storms.[20]

As a matter of historical record, who was the first to fly through a hurricane? Officially the honor has been given to Colonel Duckworth for his 27 July 1943 flight. Some historians mention that a Mrs. Dot Lemon claimed to have made the first flight through a hurricane in a Stinson Reliant in 1932.[21] No details were given and research has failed to find an account of this flight. At a meeting of the Air Weather Reconnaissance Association, in Tucson, Arizona, in October 1994, Air Force recon pioneer Edgar Crumpacker, Camp Sherman, Oregon, asked me, "Do you want to know who flew the first hurricane?" "Yes," I did want to know. "It was

Lt. Col. James B. Baker," Crumpacker said. "Call him—he lives in Amarillo (Texas)."

In 1942, Colonel Baker operated weather stations along the South American upper coast, from British Guyana to Natal, Brazil. Hundreds of young inexperienced air crews were flying along this "aerial highway," taking the long frightful trip to Africa and India, and the losses were about one in ten crews. A concerned Colonel Baker wanted to fly into a hurricane since he was anxious to know more about the weather that threatened his transient crews. So, in June 1942, Baker flew a B-18 into a hurricane off the coast of Puerto Rico, and he believes this was the first flight into a tropical cyclone. He found the same weather other crews found later: heavy rain, severe turbulence and a distinct hurricane formation although he did not find the eye as well defined as others have seen it. "I have flown in worse conditions in fronts and thunder storms," Baker said.[22]

Many other stories were told, before and in early World War II, about military and airline crews "accidentally" flying into the edge of Caribbean tropical storms. Army flight regulations strictly prohibited flying into hurricanes, but the "forbidden fruit" lure caused crews to "stray off course" and find themselves in a hurricane for much the same reason that Duckworth flew his Gulf storm. In the storms, pilots reported excitement, rough flying and apprehension, but none reported the storms to be dangerous. Capt. Gordon H. MacDougall, Boyce, Virginia, weather officer at Coolidge Field, Antigua, made two flights into a storm gaining strength in the Caribbean Sea on 21 and 22 August 1943. As the storm approached Antigua, forecaster Capt. Hugh W. Ellsaesser arranged for a transient B-25 to "fly out and look around."

The B-25 crew and Captain MacDougall headed off with no chance to get acquainted or plan strategy. Communication consisted of shouting in the pilot's ear. The pilot ignored directions of the weather forecaster as he skirted the edge of the storm. There was no rain, no turbulence, but everyone could see the huge storm off to the left. However, on return, the pilot told everybody that there was no storm, and explained that he did not want to get too close to the point of no return from Antigua in case he had to abort the mission.

In spite of the pilot's denial, rumor spread quickly that there was a storm east of the island. Pressure kept dropping, with a steady backing of the wind and a long-period swell that could only be associated with a major disturbance. Captain MacDougall explained that on 22 September, the last plane on the island was a B-18 and the only pilot was Col. Robert Alan, base commander, who was interested in the storm and wanted to take a flight. "So Colonel Alan, Major Hutchins, executive officer from a nearby infantry base, and I took off through the overcast and climbed to about eight thousand feet to get on top. To the east northeast was a great pile of dark clouds. The clouds marked the center of the storm and we flew toward it for an hour, at 170 MPH. [We were] in the overcast soon after we started with moderate turbulence but no rain. We dropped to four hundred feet [and] got our first look at the sea—a frightening sight—with tremendous waves whose tops were blown off in white sheets of water [in a] raging white spray. We bumped along at 150 MPH, drifting to the right at some incredible rate. For those of us who had spent enough time in the Caribbean, it was hard to believe what we saw: crests in long swirls blown by a wind exceeding 70 MPH. with long streaks of foam streaming from one wave to another. We had not hit the eye, but all on board had had enough, and we climbed back into the overcast and flew home."[23]

The B-18 was a friendly obsolete light bomber used for utility purposes by the Caribbean Defense Command. It was slow, easy to fly and extremely stable in weather. On one occasion, in 1943, a B-18 was forced to ditch near Puerto Rico. The crew was rescued by ship and the plane floated for days before it had to be shot full of holes to make it sink and not become a shipping hazard.[24]

Following the success of the 30th Weather Reconnaissance Squadron across the North Atlantic, and because of the rapid build-up of military installations along the Eastern Seaboard, the Army Air Forces began considering experimental hurricane flights. When word of the B-18 flights in the Caribbean and Colonel Duckworth's hurricane venture reached the Joint Chiefs of Staff, in Washington, they formally approved a plan on 15 February 1944 for aerial reconnaissance.[25] In April 1944, at the Columbia Army Air Base, Columbia, South Carolina, a "secret mission, involving extremely hazardous flying," was announced, with a call for volunteers. Although Army Rule No. 1 states emphatically, "Don't volunteer for nothing," dozens of men volunteered for this unknown assignment. Twenty experienced flight crewmen were selected—I was one of the "lucky" pilots accepted.[26]

After the decision of the Joint Chiefs of Staff, on 26 May 1944, the following directive came from Henry H. Arnold, Commanding General of the Army Air Forces, addressed to the commanding generals of the Air Transport Command, Air Service Command, Materiel Command, AAF Tactical Center, Proving Ground Command, Troop Carrier Command, First Air Force, Second Air Force, Third Air Force, and AAF Training Command:

> THE FOLLOWING IS FOR YOUR INFORMATION FOUR BAKER TWO FIVE DOG AIRCRAFT HAVE BEEN ASSIGNED BY COMMANDING GENERAL ARMY AIR FORCES IN ACCORDANCE WITH WAR DEPARTMENT LETTER ABLE GEORGE ZERO ZERO POINT NINE THREE OBOE BAKER DASH SUGAR

DASH ABLE ABLE FOX DASH MIKE ONE SIX MARCH ONE
NINE FOUR FOUR TO HURRICANE RECONNAISSANCE MIS-
SIONS PD . . .

Teletype transmissions (TWX), were sent in all-capital letters with
numbers spelled out and single and initial letters in "Able-Baker-Charlie"
military code. The above is the actual form of the message. "Translated,"
the following is the text of General Arnold's directive:

> The following is for your information. Four B-25D aircraft have
> been assigned by Commanding General Army Air Forces in accor-
> dance with War Department letter AG000.930B-S-AAF-M 16 March
> 1944 to hurricane reconnaissance missions. These aircraft will oper-
> ate during season 1 June to 1 December 1944 from Army Air Bases in
> Domestic United States or Caribbean area. Flights will be directed by
> Commanding Officer Army Air Forces Weather Wing and will be inter-
> mittent depending upon number, intensity, and location of storm
> centers. It may be necessary for these aircraft to operate between
> bases in domestic United States and bases in Bermuda or Caribbean
> area. Routine clearances will be obtained for these flights from depar-
> ture airfield. Changes in flight plan while in flight will be filed by ra-
> dio to nearest flight control or Army Airways Communications Sys-
> tem radio facility. Serial numbers of aircraft so flown are: 41-30304,
> 41-30305, 41-30449; and 41-30465. Pilots for each of the above air-
> craft, but not respectively, are Captain Allan C. Wiggins, Serial
> 0403461, First Lieutenant Otha C. Spencer, Serial 0664679, First
> Lieutenant Harry H. Helvenston, Serial 0730906, and Second Lieu-
> tenant Fred J. Huber, Serial 0535746. All MOS numbers 1081. Each
> flight of above numbered aircraft will be specifically ordered by Com-
> manding Officer Army Air Forces Weather Wing or his designated
> representative. Your command is relieved of all responsibility for safe
> operation in flight of these aircraft when so ordered.
>
> Arnold[27]

Gen. George C. Marshall, Army Chief of Staff, sent the same directive
to the commanding generals of Eastern Defense Command, Southern
Defense Command, Caribbean Defense Command and the commanding
officer of the Bermuda Base Command.

Unknown at the time, our crews would become participants in a new
weather recon project and pioneers in hurricane reconnaissance. Our orders read, "asgmt to AAF Wea Wng, Asheville, N.C. and overseas asgmt."[28]
Organized into four B-25 crews, we were members of a new experimental
squadron, called the "Army Hurricane Reconnaissance Unit." Our mission was to perform reconnaissance in tropical storms and hurricanes.

The B-25 Mitchell had proved so successful along the North and South
Atlantic ferrying routes that it was selected for the hurricane unit. The

four new B-25Ds were equipped with a complete set of weather instruments, automatic pilot, two radio altimeters, standard driftmeter and AN/APQ-13 cockpit radar. Each crew consisted of pilot and copilot, navigator, weather officer, flight engineer, radio operator and a crew chief who remained on the ground. At this point in weather reconnaissance, there was no designation of weather aircraft, such as WB-25—it was simply a B-25D. The aircraft had all armor plate and armament removed and a 960-gallon fuel tank installed in the bomb bay. Cruising speed was about 220 mph, with a maneuverable ceiling of twenty thousand feet and a comfortable range of ten hours.

Following these directives, the first organized and dedicated army hurricane reconnaissance unit was put into operation. Hurricane recon Crew No. 1 was piloted by Capt. Allen C. Wiggins, and I was a pilot of Crew No. 2. We were first stationed at Air Transport Command's (ATC) 36th Street Army Air Base, Miami, but after 1 June, we were transferred to Morrison Field, West Palm Beach, Florida. Crews No. 3 and No. 4, piloted by Lts. Harry Helvenston and Fred Huber, were stationed at Borenquen Field, Puerto Rico. (Borenquen was later renamed Ramey Air Force Base.) Missions for the four crews were ordered by Grady Norton, Director of the National Hurricane Center (NHC) at Miami, where hurricane tracking was done and storm advisories issued. Inflight weather observations were sent to Maj. Irving Porush of the hurricane center. Crews had no special training for the missions, except that first pilots were required to have one thousand hours of pilot time and two pilots were sent to the 30th WRS for one month in the spring to fly as copilots of B-25 crews flying along the North Atlantic.

Because of the historical significance of being a part of the first organized hurricane reconnaissance crews, in addition to the pilots, the other crew members included Crew No. 1, Allen M. Priester, copilot; Redding W. Bunting, navigator; Dwight E. Day, flight engineer; Frederick J. Paquin, radio operator; Crew No. 2, Arthur J. Lincks, copilot; John C. Bortz, navigator; John E. Terrell, flight engineer; Raymond B. Merritt, radio operator; Michael DeZazzo, crew chief and George W. Ebersold, weather officer; Crew No. 3, Lewis C. Perry, copilot; Billy B. Booth, navigator; Erle T. MacDonald, flight engineer; Richard W. Getchell, radio operator; Andrew Hamilton, crew chief and Raymond F. McNeil, weather officer; Crew No. 4, William J. Gillman, copilot; Kenneth Ray, navigator; Thomas J. Pousson, flight engineer; Earle W. Watson, radio operator and Edward Higley, crew chief.[29]

With the hurricane recon crews waiting impatiently at Morrison and Borenquen, the season, which officially began 1 June, started quickly. The first weak disturbance was tracked from 15 to 18 June, north of Cuba, and this was a reconnaissance orientation flight for the two pilots stationed in

Morrison. The mission was to Puerto Barrios, Guatemala, with landing at Batista Field, Cuba. Meanwhile Lts. Huber and Helvenston, at Borenquen, were searching areas of the South Atlantic north and east of Puerto Rico. The four recon crews in their new B-25Ds were absolute newcomers to hurricane reconnaissance and not knowing the habits of tropical storms, they were crews searching for a problem. The second disturbance crossed the Gulf of Mexico with no definite circulation and moved into Mexico and was not included in the 1944 list of storms. Hurricane crews were subject to day or night call, and any report of a suspicious area, such as a tropical depression or west wind from a ship or aircraft, would alert one of the crews for a mission. In 1944, there were ten disturbances identified as easterly waves or tropical storms, but only four became full-blown hurricanes.[30]

On 8 September, a real howling storm that would earn its name as "The Great Atlantic Hurricane of 1944," was first discovered by Lts. Huber and Helvenston when it was only an easterly wave northeast of Puerto Rico. This tropical disturbance would grow into a great hurricane and become the ultimate teacher and test the flying skill of these first recon crews and the mechanical strength of the B-25D. As the storm slowly moved toward Florida, Huber and Helvenston penetrated its core and reported extreme winds, heavy rain and severe turbulence. As the hurricane moved north, the two Borenquen flights set up operations in Miami for further assaults on the giant storm. These were the first official penetrations of a major hurricane by Army Air Force reconnaissance flights.

For 2nd Lt. Fred Huber's September 10 flight, Milton Sosin, correspondent for the *Miami Daily News*, received permission to fly into the storm with the recon crew for a first-hand report. As the reporter met the hurricane crew in operations, he was disturbed to see that Huber was only a second lieutenant. Sosin quietly asked the operations officer if he thought it would be safe to fly with a second lieutenant. The operations officer was firm, "If Huber can't get you back, nobody can. He has more flight time and experience than any pilot on this base."[31] Sosin flew the mission and filed this report,

> I went hurricane hunting Wednesday with the Army's Hurricane Reconnaissance unit. We found the hurricane undoubtedly! For more than two hours, 2nd Lt. Frederick Huber, 24, of Johnstown, Pennsylvania, flew at 5,000 feet, buffeted by hurricane winds as we headed into the storm. We found the hurricane 400 miles northeast of Palm Beach—a savage 100-mile-an-hour gale accompanied by a torrential downpour so heavy that streams of water forced themselves through the plexi-glass nose of the B-25.
>
> Lt. Huber's orders from Maj. Irving Porush, from the National Hurricane Center, were to skirt the hurricane—Monday, Huber and

his crew on an earlier flight, had flown right into the center of it and Captain Allen Wiggins and another hurricane-hunting plane had also penetrated the storm. . . . Lieutenant Huber tried to obey orders, but the hurricane wasn't acting under orders. We weren't able to skirt the edges and, before we knew it, we were enveloped in the storm.

The plane pitched and tossed; bucked and swayed, but the twin engines kept roaring on. . . . Below us the ocean was streaked with greenish-gray lines . . . the waves were high . . . the plane droned on for half an hour. Suddenly we hit an area worse than any we had experienced. The nose of the ship lurched up, dropped sharply, bucking . . . the plane shuddered from side to side.

Suddenly we changed course and got the signal that hurricane hunting for the day was over—we were heading home. We streaked back to our base at 275 miles an hour with a tailwind. As we neared the coast, we came out of the cold gray nothingness . . . suddenly the sun streamed down through the glass dome.[32]

As the hurricane approached the Atlantic coast, it was time for Crew No. 2 to make a penetration—that was my crew. On the night before our scheduled mission—the fourth flight into the storm—I went to the flight line at Morrison. There was one plane on the ramp—our own B-25D, 41-30449. All other aircraft had been evacuated. Our plane had developed a propeller oil leak on previous flights, and we had had difficulty getting maintenance crews to work on it—weather recon had little priority at busy Morrison. Now, with the hurricane threatening Florida, I was pleased so see our plane bathed in lights and several civilian mechanics working. I felt much better about tomorrow's flight.

On 12 September, we took off at 0500, in light rain and gusty winds—a sure sign that the great storm was swirling several hundred miles "out there somewhere," over the open Atlantic Ocean. Our mission was to fix the storm position on that morning, reporting to the National Hurricane Center, so that its track could be forecast. Crew No. 1, piloted by Captain Wiggins, would take off at noon from Bermuda and fly through the storm from the east.

Our plan was to "box" the storm—to try to fly around its counter-clockwise circulation. To circle the storm, we would fly east until the wind was off our left wing (from the north), then fly south until we had a west wind; then fly east until a south wind came, then turn north until we had an east wind—then head for home, hopefully through the "eye" of the storm. This box pattern would provide tail winds most of the time, and we could circle the hurricane more quickly.

We kept position by taking radio fixes and transmitting RF signals. Flight altitude was from five hundred feet to ten thousand feet for pressure and temperature readings. At five hundred feet we could see the ocean and found the wind speed to be about 110 MPH—a radio altimeter helped maintain an exact altitude since the pressure altimeter, because of the low pressure near the center of the storm, was not accurate enough for safety. Cockpit radar helped find soft spots in the clouds so we could avoid the extreme turbulence. The rain was heavy—heavy!

For a hurricane moving north, the southwest quadrant (the trailing edge) is the most comfortable to fly and the northeast quadrant (the leading edge) is the most treacherous. For example, if a storm with 110 MPH winds, is moving at fifteen miles per hour, the winds in the southwest quadrant would be 95 MPH (110 minus 15) and the winds in the northeast, leading edge quadrant would be 125 MPH (110 plus 15). Our greatest anxiety in an Atlantic storm was on the east side because there we were the greatest distance from home base, with a storm between us and safety, and there are no alternate landing fields in that part of the Atlantic. With so little hurricane experience, we felt like Columbus' sailors about to drop off the end of the world. And I remembered Columbus' words as he wrote in his log, after an encounter with a Caribbean hurricane, "Nothing but the service of God and the extension of the monarchy would induce me to expose myself to such dangers."

On the backside of our flight plan, Lt. George Abersold, weather officer, decided the storm was too large to fly around. We had fought heavy rain and turbulence for several hours and I, for one, was ready to find the eye and go home. But Lt. John Bortz, navigator, and Abersold, weather officer, decided we should hold our heading for six more minutes. We would then turn to 270° and try to find the eye. By this time, the hurricane was really concentrating its fury on us—at times we thought the plane would be torn apart from the turbulence. We were soaked; the cockpit was drenched, the engine cylinder-head temperature was dropping, and we were afraid the heavy rain might drown the engines. We closed the air scoops and put the fuel mixture at full rich and the Wright-Cyclone engines kept running smoothly. Copilot Arthur Lincks and I fought to hold the B-25 in level flight. I looked with a question at the navigator. "Five more minutes," he said. Ironically, just at that instant, one of the engines backfired slightly, caught and roared on. The navigator looked at the weather officer, received a slight nod, and turned to me, saying, "Turn to 270, now!" Although we were headed into the mighty ferocity of the hurricane, we were on our way home to the Florida coast!

Instrument flight requires absolute attention—in heavy turbulence, gyro instruments are useless—the magnetic compass oscillates wildly and the most reliable instruments are the basic needle-ball, altimeter, rate of climb and air speed. Suddenly, after a jolt through the wall cloud, we found a clear eye of the storm, with scud-type clouds near the ocean. The sky was clear, the sun was shining brightly through high thin clouds—the air was smooth and flight conditions were perfect. After finding and reporting the center of the storm, our mission was complete. Now as we were headed home, I experienced what would become one of the most vivid memories of this flight. After the darkness of heavy clouds, the cockpit became radiant with light—the sun was shining through a misty rain, and for the first time I felt sure we would make it.

We landed after a six-hour and thirty-minute flight, which included an hour and ten minutes of rainy, turbulent, pre-dawn darkness, and three hours of instrument flying. The crew was soaked and, after landing, the plane dripped for hours on the ramp. The rain had stripped all of the paint from the propellers. Later that day, Captain Wiggins' Crew No. 1 flew through the treacherous northeast quadrant of the storm and reported flight conditions from Bermuda much worse than we had encountered. Wiggins' plane lost 110 rivets in the nose and the paint was stripped from the leading edge of the wings.

The Great Atlantic Hurricane turned northwest, missing Florida, developed winds up to 140 miles per hour, and tore its way up the heavily-populated eastern coast. Experiencing a hurricane, as it storms through city and village, can be high drama—"weather gone mad!" But the usual descriptions are not exaggerations—they are too true. For example, as the winds swirled through metropolitan New York City, *Life* magazine described the 1944 Great Atlantic Hurricane: ". . . a black funnel of a storm with a 140-mile-an-hour gale circling its core of calm . . . hit New Jersey, Long Island and the New England coast. The next day, in pale sunlight . . . storm-struck inhabitants began counting the hurricane's toll. At least twenty-seven people had been killed, one Navy destroyer and two Coast Guard ships sunk, and $50,000,000 in damage. . . . Thousands of homes deprived of water, telephones, electricity. . . . Whole houses toppled into the water with great elm and pine trees fallen."[33] At Atlantic City, the famed boardwalk was ripped up and smashed through the steel pier [as] the hurricane moaned between skyscrapers in 95-mile-an-hour gusts. . . . Thousands of people stayed in downtown buildings, watching the storm crashing through the stone canyons [as] water crept into the subways [and] trains stalled. . . . Then, state by state, the wind died, the rain stopped and the stars came out. This great hurricane did one fifth the damage and claimed less than one tenth the lives lost in the *Grandmother* storm of 1938. Storm warnings had served their purpose."[34]

The B-25 was well cast for its roles in World War II and was the grandparent of all weather recon planes. I might not be living today if the B-25 had been less of an airplane. One small example: On one evening, in August 1944, after a storm mission that had been more messy than turbulent, we were landing on a 1500-ft runway in Antigua. Rain greased the strip and after touching on the first foot of runway for a short-field landing, I applied brakes; the plane went into a skid and we were headed for the Atlantic Ocean. I screamed, "Turn, dammit," cut the left engine throttle and gave full power to the right engine. The B-25 shuttered, made a graceful pirouette on one wheel and headed 180° back down the runway with the ocean behind us. My legs turned to jelly and I could not control the brakes for steering—the copilot taxied to the flight line.

In the same storm, Lt. Fred Huber and his crew were coming into Kingston, Jamaica, from a mission that started in Puerto Rico earlier in the day. Clouds and a foggy steamy mist shrouded the airfield—the men were tired and looked forward to a dip in the Caribbean. As Huber was preparing to land, bouncing just below the two hundred-foot ceiling, Kingston tower called that a Navy blimp was down over the ocean south of Jamaica. The tower control officer asked, "Could they fly out and try to find it?" With verbal clearance, Huber pulled up wheels and headed south, flying fifty to one hundred feet off the water. Visibility was almost zero. In one of flight's rare miracles, they immediately found the downed blimp in the fog—like a huge cigar sticking out from the ocean. After sending an RF signal, Huber told Alan Priester, his copilot, that he would circle around so he could take a picture. After a tight 180° turn, they lost the blimp and were never able to find it again. Flying back to the field, they landed, Navy air-sea-rescue picked up the blimp crew members and Huber and his men got their dip in the surf.[35]

After the 1944 hurricane season, when the storms had given way to winter, Col. John K. Arnold, Regional Control Officer of the Ninth Weather Region, wrote Col. W. O. Senter, Commanding Officer of the AAF Weather Wing, Asheville, North Carolina,

> It is most gratifying to recognize . . . the assignment of crews . . . on 18 May 1944 for . . . reconnaissance in tropical storms and hurricanes, . . . an entirely new operation . . . of flying, weather reporting and forecasting. . . . Personnel assigned to the Army Hurricane Reconnaissance Unit have operated in the Caribbean, Gulf of Mexico and Western Atlantic during the 1944 hurricane season from 15 June to 1 December. Reconnaissance missions were flown into all tropical disturbances, storms and hurricanes over water areas where no other weather data was available.[36]

Commendations were also received from F. W. Reichelderfer, Chief, Weather Bureau, U. S. Department of Commerce; Jesse Jones, Secretary of Commerce, and Henry L. Stimson, Secretary of War.[37] This documentation is included because, in the history of weather reconnaissance, the 1944 season has been a vague blank, when in fact, 1944 was the real beginning of organized hurricane reconnaissance.

Although a hurricane or typhoon can form at any place in warm equatorial waters, between 60 and 70 percent of hurricanes begin as "easterly waves" off the northwest coast of Africa and move into the doldrums of the South Atlantic. This area is the roaming grounds of the nebulous "intertropical convergence zone" (equatorial front) as it moves above the equator to the warm calm waters of the Atlantic or Caribbean, causing massive thunderstorms and cumulus buildups above fifty thousand feet. As the equatorial front wanders above and below the equator, at the whim of upper air movements and pressures, there is a conflict of southern hemisphere clockwise and northern hemisphere counterclockwise rotation around low pressure systems. An intruding polar trough causes a wave, and the low pressure area brings in warm moist air to create a west wind and a counterclockwise rotation. As the air is heated, it rises as surrounding moist air rushes in to take its place, intensifying the circulation and, in time, a hurricane can be born. The rotation of the earth, pushes the storm into a slow movement toward the west, curving slightly to the north. In most cases, the vortex would die and no storm would form although near the time of the autumnal equinox in the Caribbean, when the heat has attained its peak, one wave may occasionally become a monster of violence.[38]

Before World War II, weather officers were trained in "stateside" frontal weather and had little knowledge of tropical meteorology. To learn about local weather, many weather officers had been told to "talk to the natives," to find out how they knew when a storm was threatening. Yet, few young weather officers, with degrees from such prestigious schools as Massachusetts Institute of Technology or California Tech, would bother to talk weather with natives. However, one weather officer, from Hurricane Crew No. 2, was 1st Lt. George Abersold, whose favorite pastime was drinking rum in the taverns of Puerto Rico or Cuba, hanging out with the local sages, asking, "How do you know when a storm is coming?" This question always created excited interest and he learned some non-textbook folk meteorology.[39]

There were many secrets known only to weatherwise natives who had spent their lives in fear of the hurricane winds and could detect approaching storms still hundreds of miles out in the open ocean. Base commanders found that, long before a storm was detected by weather instruments or forecasters, the absentee rate of local workers increased, and they knew that out in the deep ocean a storm was building. No one knew exactly how older natives knew the secret "laws of the storm." It seemed as if ancient instincts were whispering to them that the deadly winds were headed their way. High, wispy cirrus clouds may have been a sign, but a slowing of wave frequency and motion on the beaches was the real "teller of secrets," long before storms became a threat. Other indicators were a slow, steady drop in air pressure and a rise in temperature. Being island people, the natives knew that waves hit the beaches at an average of eight per minute. When the frequency dropped to five or four, the islanders knew a real storm was on its way. If the waves continued in one direction, the storm could be expected to hit head on; a change in direction indicated that it would pass to one side.[40]

What is it like to fly into a hurricane? In 1944 it was a new experience for military crews, and no one knew what to expect. Hurricanes were as different as they were erratic and as frightening as the ancient lore of weather could predict. Yet, after several penetrations, most crews felt safer in the air than on the ground. There was rain—drenching rain—that penetrated through cracks and seams of most aircraft, soaking the crews. There were lightning, thunder, and light hail. The ultimate shock came when crews dropped down to take surface readings and below the plane just a few hundred feet swirled the terrible anger of black water with long horizontal sheets of spray streaking for hundreds of feet. To see the violent water was numbing. Never was there a member of a hurricane crew who didn't imagine, with reverent horror, plunging into those waters, and never was there one who didn't pray that the engines didn't quit and the wings stayed on. Through this experience of horror and fear, crews talked of a poetic experience—a special awe—at sharing one of nature's wildest moods.

Col. Albert D. Purvis, a commander of the 53rd *"Hurricane Hunters,"* the most experienced of all squadrons, probably expressed the feeling of every pilot about to fly into a storm,

> There is always fear whenever anyone flies into a hurricane . . . keeping you on edge, alert, cautious. The most dangerous part is flying into the wall of thunderstorms that rings the center of a hurricane. . . . As we penetrate that wall . . . turbulence increases, rain spews against the skin with machine-gun intensity . . . like you are inside a popcorn popper. . . . All of a sudden it is quiet . . . except for the drone of the engines. . . . it's so quiet, its spooky. We've hit the eye of the storm . . . the serenity and beauty within the center will literally cause your jaw to drop in reverence. In the middle of nature's fiercest storm with white, ice-like clouds towering above you is a crystal-blue hole in a shrouded sky [and] the churning white and green waters calm into an almost rippleless surface at the center of the storm. . . . Fear and anxiety leave until you realize that you have to fly back into the wall of thunderstorms in a precious few minutes.[41]

The danger and sheer violence are immediately forgotten as the crew ponders nature's awesome performance and witnesses the beauty of the

Flying at sixty thousand feet, a U.S. Navy aircraft photographed an Atlantic hurricane. The eye of the storm and the turbulent band clouds are clearly visible. *Navy photograph courtesy of Rick Vanderpool*

storm's eye. To those lucky enough to find the center of the storm, there is the other-worldly experience of entering a perfectly calm, sometimes sunny center, often with birds flying around. In 1944, in the Atlantic, I felt spiritually blessed to witness one of life's most profound secrets, and my puny mind was unable to grasp the cosmic mystery—the depth of feeling was too great, and the hurricane experience remains one of my deeper memories.[42]

William Anderson, Boise, Idaho, who flew weather recon for six years, describes his hurricane experience in a night flight. "The eye of the hurricane is nature's sleight-of-hand that must surely have proposed perplexing questions for primitive people. First comes the violence of the storm, then the tranquillity of the eye, followed by another crashing fury of winds from the opposite direction—to smash through its forbidding wall is to come upon an unbelievable spectacle witnessed by only a handful of brave, foolhardy souls—a full moon shining in intermittent flashes through galleried cotton clouds in the eye's amphitheater, the star-speckled sky, the languid sea beneath is to become a Conway, finding his Shangri-La."[43] Tom Robison, Ossian, Indiana, passenger on a weather recon flight out of Guam said, "The worst problem was trying to decide if the feeling in my stomach was an emptiness wanting to be filled or a fullness wanting to be emptied."[44]

For hundreds of years, hurricanes in the West Indies were named after a particular saint's day on which the storm occurred. For example, there was "Hurricane Santa Ana," which struck Puerto Rico on 16 July 1825, and "San Felipe the first," and "San Felipe the second," which hit Puerto Rico in 1876 and 1928. An Australian meteorologist, Clement Wragge, began giving women's names to tropical storms before the end of the nineteenth century. During World War II, although not official, this practice was widespread by Army and Navy meteorologists plotting typhoons over the Pacific.

To prevent confusion if there were two storms at the same time, in 1947, the Weather Bureau, as well as Air Force, and Navy meteorologists, began naming storms using Able, Baker, Charlie, etc. After two years, this naming was abandoned, and storms were given female names, such as Alice, Brenda, Connie, etc. Experience showed that short distinctive first names were less confusing than the longitude-latitude identification methods. The Weather Bureau received a great many letters from female writers who heartily endorsed identifying hurricanes with women's names. Lt. Edgar W. Donaldson, B-50 pilot of the 53rd, commented, "Storms are named for women for a damned good reason. No two are alike and no two are the same twice. You may go into one in the morning and it's like a piece of cake. You go back at night and you're lucky to get out alive."[45] This practice of using solely women's names ended in 1978 for storms in the Northeastern Pacific and, in 1979, alternating male and female names were used for Atlantic, Caribbean and Gulf storms. The names of specific individuals are not chosen for hurricane names.

A six-year name list, with male, female, and names of international flavor, was adopted by the Weather Bureau in 1991. If a tropical disturbance develops a counter-clockwise rotary circulation and winds build above thirty-nine miles an hour, it will receive a name from the list. After six years, the names will be repeated. For example, here are the 1997 proposed names, although occasionally a name is changed: Ana, Bob, Cindy, Danny, Erika, Fabian, Grace, Henri, Isabel, Juan, Kate, Larry, Mindy,

Nicholas, Odette, Peter, Rose, Sam, Teresa, Victor and Wanda. Names for especially destructive storms are retired from the six-year cycle.[46]

In the Pacific Ocean a similar but separate system is used to identify tropical cyclones of typhoon intensity. Because the Pacific has a greater number of tropical storms each year, four sets of twenty-one different girls' names are used, in a continuous fixed sequence without regard to the year or season. The first typhoon in the Pacific during each season is assigned the name following the last name used during the previous season. When all eighty-four names have been used, the entire Pacific list is repeated, starting with the first name in the first set.[47]

When the 1944 tropical cyclone season was concluded, the Hurricane Reconnaissance Unit was disbanded and the four crews sent to the Tenth Weather Squadron at Gushkara/Barrackpore, India, for reconnaissance over the Bay of Bengal and the China-Burma-India (CBI) Hump. To replace the 1944 hurricane unit, in May 1945, in time for the Atlantic hurricane season, "Duck Flight," a group of five B-25 crews of the 1st Reconnaissance Squadron, moved from British Guyana to Morrison Field for weather duty.

During the 1945 season, the crews of the 1st WRS followed one of the most violent storms of the decade. It was "Kappler's Hurricane,"[48] named after 2nd Lt. Bernard J. Kappler, weather officer on the B-25 crew piloted by 1st Lt. Dennis A. Cassidy, from Peoria, Arizona. The storm had formed over Western Africa and continued its dramatic movement across the Cape Verde Islands, but was then lost far out in the Atlantic. Kappler discovered the storm again on 12 September 1945 on a regular recon flight near the Windward Islands. It was tracked every day by one or more of the B-25s. "Kappler's Hurricane" was also the first hurricane to be penetrated by a B-17 of the newly organized 53rd WRS. The B-17 was a new long range, four-engine aircraft being used for hurricane reconnaissance to replace the B-25.

What made "Kappler's Hurricane" famous was that its center crossed the tip of Florida, and the storm moved up the peninsula, paralleling the two coasts and ripping the heartland out of Florida. Leaving Florida, the hurricane skirted the coast of Georgia and moved into South Carolina. Lt. Edward Bourdet, weather officer on the last flight into Kappler's storm, wrote: "The worst part of flying hurricanes is if there should be trouble that would force the plane down, the crew would have no chance of getting out alive. The best part is that you are instrumental in providing adequate warning, . . . saving lives and property."[49] This theme was restated many times by hurricane reconnaissance crew members. War was the process of killing and trying to keep from being killed. Weather recon, even in wartime in military aircraft, was an exercise in trying to get through the storms alive, to provide a warning and a service to people.

Weather recon during the bitter North Atlantic winters was considered too hazardous for twin-engine aircraft, and the weather service wanted the security and range of four-engine aircraft. While the B-25Ds served well in short-range situations, four engines are better than two, and, after the 1943 season, the 30th WRS asked again, "Could, perhaps, the Army spare three B-17s?" The Army agreed and in December 1943, 30th WRS crews were sent to Spokane, Washington to pick up three "new" B-17s. What they found were three 1941, war-weary veterans of many Aleutian missions. The planes were in such sorry shape, it was February 1944 before they could be put into operational condition.[50] Lt. Col. Don Offerman, New Braunfels, Texas, pilot for the 53rd WRS, wrote in his memoirs, "Air weather had a low priority for equipment."[51] David Magilavy, president of the Air Weather Reconnaissance Association and twenty-year veteran of weather recon and equipment development, complained, "Weather had a low priority for everything and a constant problem was trying to keep the hand-me-down weather aircraft in operation. Parts were difficult to get and many aircraft were kept in the air by scrounging and horse-trading practices, survival techniques learned by maintenance personnel."[52]

The 3rd Weather Reconnaissance Squadron, Medium, was activated at Grenier Field, New Hampshire on 31 August 1943. On 15 June 1945, the 3rd was Redesignated the 53rd Reconnaissance Squadron, Weather, Heavy, and, in December 1945, when World War II was over, the 53rd inherited the planes, personnel and mission of the old North Atlantic 3rd WRS. This reorganization completed the direct lineage from "day one" of the nation's military weather recon force. Lt. Col. Karl T. Rauk, La Crescent, Minnesota, was commander of the 3rd, and Maj. George L. Newton took command of the newly activated 53rd. The 53rd became responsible for all weather reconnaissance over the North and South Atlantic, the Caribbean and the Gulf of Mexico.[53]

A world-wide reconnaissance net for weather warnings was being put in place. In 1945, in addition to critical Atlantic and Caribbean hurricane missions, three flights of the 53rd were deployed to other areas where weather started its movement to the south. Flight Alpha flew synoptic tracks out of McCord Field, Washington, keeping watch over the North Pacific and the Gulf of Alaska to intercept Arctic storms approaching the Northwest. Flight Bravo patrolled the North Atlantic out of Gander, Newfoundland; Goose Bay, Labrador, and Grenier Field, New Hampshire, scouting frigid blasts from weather birthing areas of the Arctic. Flight Charlie operated out of Lajes Field, Azores, to patrol hurricanes and storms threatening Britain and Europe. In November 1946, the 53rd was moved to Morrison Field, Florida, while still patrolling the ferry routes over the North Atlantic and hurricanes in the South Atlantic. There the 53rd earned its name *"Hurricane Hunters."* Later, in June and July 1947, the 53rd was deployed to Bermuda and became a part of the 308th Reconnaissance Group, Weather.[54]

After the war, thousands of surplus bombers became available and the B-29 was the favored new aircraft for weather recon. With better equipment and more weather experience, crews began to snoop around in areas of the storms that had been impossible with smaller aircraft. The B-29 made new accomplishments and better reconnaissance possible. On 7 October 1946, Maj. Paul E. Fackler, pilot of a 53rd WRS B-29, became the first to fly into the top of a hurricane. Flying from Bermuda, Fackler's B-29 climbed to twenty-thousand feet and headed toward the storm center, about 350 miles to the southwest. At thirty-one thousand feet, the crew went on instruments and began probing the storm with radar, and penetrated the dangerous northeast quadrant with winds over one hundred miles per hour. By radar they found ring-like clouds within the storm center, and Fackler and his crew spent two and a half hours making five penetrations. Earlier recon crews had been able to make only one penetration on a flight. Fackler estimated the top of the hurricane was thirty-six thousand feet.[55] In October 1947, the 53rd made the first low-level night penetration of a hurricane.[56]

With its beginning in 1944, tropical cyclone reconnaissance, in both hemispheres, became a major part of weather reconnaissance throughout the years following World War II. Hurricane recon history continues with several squadrons, formed after the war, specializing in storm recon throughout the world. Hurricane reconnaissance is the only part of weather reconnaissance that continues to this day, despite the dominance of satellite observations and storm tracking.

3

The Short Weather War Against Japan
1943-1945

As World War II raged in Europe, the United States was content to do its part through lend-lease and loans, in addition to its role as international cheerleader for the Allies. On 7 December 1941, Japan brought a fiery and deadly end to the U.S.'s passive role when Pearl Harbor, Hawaii, was bombed on a quiet Sunday morning—World War II had come to the American people with savage fury. On Monday, 8 December, the United States and Britain declared war against Japan, and China and India declared war against Germany, Italy and Japan. Suddenly the Far East, especially China, became important as World War II's second battlefield. Japan had been bullying China since 1931, and even though the U.S. had been providing some support to China, it was only after Pearl Harbor and an international declaration of war, that the world woke to the Japanese threat in the Orient.

China was a primitive country, three hundred thousand square miles larger than the United States, with the greatest population in the world. Vast distances, towering mountains, an undeveloped rail and highway system and a weak-willed, poorly-trained Army made ground defense almost impossible. China had an "Air Force" of fewer than seventy-five planes, with pilots who more often killed themselves than the enemy. Japan controlled most of the land and air over China and could move and bomb at will. The only way to defend China was through air power and that must come from the Allies.

Almost immediately the Allies mounted a fierce assault against Japan, with the efforts of Gen. Claire Lee Chennault's American Volunteer Group[1]—

the famed "Flying Tigers"—the British Royal Air Force in Burma, a few Chinese pilots, the Canadians, and Australians. The real reason to save China was to provide bases for the Allied bombing raids against the Japanese homeland although the all-out offensive against Japan would have to wait until the war in Europe was won.

Japan's first strategy was to isolate China from the world. The Japanese sealed off the entire Chinese coast from Manchuria to Hanoi, Indochina. Russia blocked the top of the country, and, to the west, the great Himalayan mountain range was a natural barrier from Tibet to the sea. China had no lifeline to the outside world except the Burma Road, 581 miles of winding gravel through the mountains—an open target for Japanese fighters. And, in March 1942, the Burma Road was closed and quickly reopened. But the only practical way to get supplies into China was by transport aircraft over the Himalayas, an air route commonly called the "Hump."

Building airfields and weather stations, ferrying aircraft, training crews to fight, and beginning the Hump airlift to carry the freight of war over the Himalayas took agonizingly long months. The first feeble efforts to supply China by air began in mid-1942, and the real organized effort, under the Air Transport Command (ATC), began in December the same year. The Hump supply line had unmerciful problems. Young, inexperienced Hump pilots—kids really—flying untested aircraft, and confused by uncertain leadership, paid heavily with their lives and planes. The Hump was called "The Aluminum Trail." Japanese Zeros shot the slow, unarmed transport planes out of the sky, violent weather brought them down, and planes crashed for unknown reasons. Above all, weather was the biggest problem.

Col. Richard Ellsworth, pilot and meteorologist, with weather experience in Alaska and the South Pacific, arrived in India in August 1943, as commanding officer of the newly-organized Tenth Weather Region. His plan was to establish weather stations and reconnaissance squadrons to support the ATC in its supply flights over the Hump. Ellsworth first butted head-on with General Chennault and Generalissimo Chiang Kaishek, who didn't want additional personnel in China—"more mouths to feed," they said. With weather as the greatest hindrance to the flow of supplies into China, Ellsworth finally convinced the two generals of the value of accurate weather forecasting and reconnaissance in China and began to establish weather stations along the Hump, into China, and even behind the Japanese and Communist lines both in China and Southeast Asia. Harry M. Albaugh, former reconnaissance pilot from Mariposa, California, wrote, "I feel very strongly that Gen. 'Dick' Ellsworth was the real daddy and developer of airborne weather reconnaissance."[2]

Into this brutal China-Burma-India (CBI) theater in October 1944 came the 2nd Weather Reconnaissance Squadron (WRS), Air Route Medium,

commanded by Lt. Col. James Baker and attached to Ellsworth's Tenth
Weather Region. The mission of the 2nd WRS was to fly weather route
reconnaissance over the Hump, from Barrackpore/Gushkara, India, to
Kunming, China, and synoptic flights over the Indian Ocean and the stormy
Bay of Bengal. Also in 1944, Tenth Weather organized a reconnaissance
unit in Chengtu, China, to fly weather support for proposed B-29 strikes.
Strategic planning called for Gen. Curtis LeMay's B-29s based at Kharag-
pur, Dudhkundi, and Chukulia, northwest and west of Calcutta, to strike
Indochina and targets as far south as Sumatra from the bases in India and
to bomb the Japanese mainland from the staging base in Chengtu, China.

Comfort-loving airmen of the 2nd WRS found the CBI theater an un-
tamed jungle. Living conditions in thatched bashas were primitive. Lt.
Col. Arthur McCartan, Florrisant, Missouri, who had organized the first
weather reconnaissance across the North Atlantic, remembers packs of
wild dogs and hyenas threatening personnel in the screened latrines at night—
airmen carried their .45-cal. sidearm at all times.[3] Eugene Wernette, Fort
Worth, Texas, beginning a long career in weather reconnaissance, can't forget
the rope beds, dobi itch, Atabrine tablets, dysentery, and coming home at 118
pounds. "But a tour in the CBI made pilots out of boys," he admitted.[4]

In China, Charles Markham, Boulder City, Nevada, and his crew ate
what the Chinese ate—eggs, rice, bean sprouts, water buffalo—mouthfuls
which got bigger with chewing. "Rice wine stained your teeth purple for
weeks."[5] Bob Klossner, Arlington, Texas, had been in the CBI since 1941
as an engineer to help build airfields. His memoirs include these comments: "Of all varmints in Upper Assam, (India), the krite, a small snake,
was the deadliest. If someone was bitten by a krite, send for the priest not
a doctor. Captain Pickett, from Kansas City, shot a 395-pound tiger that
measured 13-feet, 6-inches in length."[6] Yet, to the newly arrived Americans, China was a land of intrigue and mystery, as portrayed by Milton
Caniff and his "Terry and the Pirates" cartoon series, and a land of poor
and beautiful people, created by Pearl Buck in *The Good Earth*.

China was a new chapter in weather reconnaissance. Experience had
proved that synoptic flights were more valuable in forecasting than route
reconnaissance, and the most productive area for synoptic recon was the Bay
of Bengal, south from Calcutta between the mainlands of India and Burma.
These synoptic flights were flown by the 2nd WRS stationed at Gushkara and
Barrackpore, India. Although there was little military activity over the Bay
of Bengal, this region was one of the most volatile breeding grounds for the
great storms that moved into India and over the Hump, bringing tremendous
winds and monsoon rains to the western shores of Burma and into India's
Assam Valley. Official rainfall averages were 425 to five hundred inches annu-
ally, with one hundred inches each month during June, July, and August.[7]

In summer, the Hump routes were plagued by the hot Indian monsoon winds, unprecedented rain, thunderstorms, near zero visibility and icing. The world's most violent weather is in Asia, the meeting place of three turbulent air masses. Low pressure from the west, moved along the San-tsung Range, the main ridge of the Himalayas, where warm and wet high pressure systems from the Bay of Bengal clashed with cold Siberian low pressure systems. Heat rising from the jungles of Burma, intensified this violent mix of weather.[8]

Weather reconnaissance crews, flying radar-equipped B-25s, went to work quickly. From the first day the 2nd WRS arrived in India, until after the end of the war, in August 1945, weather crews made a daily flight south, one thousand miles over the Bay of Bengal, from Gushkara to the Japa-nese-held Andaman Islands, a refueling stop at Akyab, Burma, and return to Gushkara. Flights were also scheduled over the Indian Ocean and Indo-china, although not on a daily basis. Montgomery Truss, meteorologist-gunner, from Birmingham, Alabama, described these flights, in aircraft that were ". . . fully armed, with extra fuel in bomb bay tanks for the ten-hour missions. There were severe storms over the Bay, monsoons and ic-ing and turbulence beginning their movement over the Hump mountains. Monsoon storms, south of Rangoon, had winds exceeding seventy-five miles per hour and the flight was always rough."[9]

Weather observations, with drift meter wind measurements, were made every thirty minutes from five hundred and ten thousand feet and sent to ground stations. Meteorologists on recon flights also checked the coast line for any significant ground swell which would indicate tropical cyclones farther out in the Bay of Bengal or in the Indian Ocean. In 1944 and 1945 there were no Japanese in Burma although occasionally fighters from the Andaman Islands would challenge the B-25 crews. When "jumped" by Japanese fighters, the recon planes simply took cover in the clouds and the fighters could not follow. Mechanical problems were com-mon. On one mission, crews of a returning B-25 could not get the nose wheel to drop. The pilot radioed his problem to Gushkara, gave instruc-tions, and prepared to land. With main wheels lowered, he had the entire crew go to the tail. As he landed nose high, a jeep with two men on the hood caught the plane's tail as it slowed on the landing roll and, with the weight of the crew in the plane, kept the tail on the ground. This maneu-ver saved a valuable aircraft and possible injuries.[10]

Problems encountered by the ATC flying through the storms over the Hump and inaccuracies in forecasting raised the ire of Col. Thomas Hardin, expert pilot and hard-driving ATC commander. Colonel Hardin earned his place in CBI history when he issued a decree to his pilots that, "There is no weather over the Hump," meaning that crews could not cancel flights or turn back because of weather.[11] Hardin wired the India-China Wing commander and Tenth Weather commander Colonel Ellsworth that "the weather service is wholly unsatisfactory, . . . the only thing worse than communications." Ellsworth blamed an improper statement of Hump weather needs and shortages of men and equipment.[12]

The ATC, as principal user of weather information, tried to take over the weather service, but was not successful.[13] Because of constant complaints, the historian of the Tenth Weather Squadron, as he evaluated the CBI, wrote, "Weather was a small tadpole in a big pond of croaking frogs."[14] Weathermen were "affectionately" known as "balloon blowers" with all of the prestige that went with the title. However, Ellsworth and Hardin, in an act of mutual self preservation, began to work together, and ATC finally decided that, "in a weather factory as volatile as the CBI," forecasting by the Tenth was satisfactory.[15] The two former antagonists even worked together to pioneer night flights over the Hump.[16]

However, Generalissimo Chiang and General Chennault continued to object to weather squadrons increasing their personnel in China. Since no food was being flown over the Hump, the Chinese would have to grow more rice to feed them.[17] Although the Chinese did provide food for the new personnel, Tenth Weather was required to supply its own stations and transport all weather personnel without increasing the ATC burden. Pilots of the 2nd WRS, when they flew weather recon over the Hump, always carried a full load of supplies, mail, replacement personnel, etc. Weather crews flew transport duty as well as reconnaissance.

In 1944, Gen. William Tunner came to the Hump as ATC commander to try to reduce the terrible losses. As he personally flew the Hump, he experienced weather changes from minute-to-minute and mile-by-mile, from "the low steamy jungles of India, to the mile-high plateau of western China . . . thunderstorms . . . icing . . . turbulence greater than I have ever seen elsewhere in the world. Winds as much as a hundred miles an hour, would glance off the mountains to create updrafts over the ridges and down drafts over the valleys. Planes would drop five thousand feet a minute, then suddenly be whisked upward at the same speed."[18]

In addition to weather problems, the CBI theater was possibly the most politically confused and disorganized theater of the war. The CBI seems to have gathered the strangest assortment of cantankerous characters ever assembled in one theater. Foremost was Madame Chiang Kai-shek, the "tiger lady" who wielded unmerciful political influence with the President and the U.S. Congress. She was educated at Wellesley College in the United States, spoke fluent, beautiful English, and was ready to lay down anyone's life for her China. With Madame Chiang was her Methodist husband, Generalissimo Chiang Kai-shek, whose main enemies were the Chinese communists and not Japan. The third antagonist was Gen. "Vinegar Joe" Stilwell, who spoke fluent Chinese, had the confidence of the Chinese armies, and was proving to be one of the true military geniuses of World War II. Stilwell believed that war on the ground was the only way to defeat Japan. His sometimes antagonist, Gen. Claire Lee Chennault, another military genius, was an outspoken evangelist who preached and perhaps proved that only air power would be able to defeat the Japanese.

Before the B-29s came to the CBI, Col. Richard Ellsworth, in November 1943, had proposed a China Weather Central, a network of weather stations and recon bases in China, to provide weather forecasts for B-29 bombing. With twenty-five officers and seventy-five enlisted men, China Weather Central was to be located near Chengtu, outside of the village of Hsingching. The Chengtu area was selected because Russian broadcasts could be easily picked up, Pacific broadcasts came in strong and few American troops were in the area. This plan was approved by Gen. George Stratemeyer and Gen. Claire Chennault.[19] Later the closely guarded secret of the B-29 staging base was announced, and it too was located at Chengtu. Suddenly, the new weather unit found itself in the middle of the Twentieth Bomber Command, with a B-29 airfield under construction outside its hostel gate. This coincidence caused many officials to believe that there had been a serious leak in their plans or else Colonel Ellsworth had a prophetic gift. He denied both and said that Chengtu was the only logical place in all of China for Weather Central. The Army must have reached the same conclusion for the B-29 airfield.[20] When the field was completed, Weather Central was ready for the *Superfortresses* and their first bombing mission to Japan.

The real reason weather reconnaissance had been brought to the CBI was to support B-29 missions against Japan. Synoptic recon observations over the Bay of Bengal and the Indian Ocean helped ground weather stations provide accurate forecasts. The B-29 *Superfortresses* came to India and China in April 1944, following the build up of weather stations and establishment of reconnaissance routes in China. Just two years after Col. Jimmy Doolittle's B-25s bombed Tokyo, the huge B-29 bombers were ready to carry the war to Japan. The unimaginable distances, mountains and weather presented an almost impossible tactical problem. To cover the vast distances across India, Burma, and China to Japan, bombing missions were planned in four stages. From bases near Calcutta, the B-29s would fly across the Hump to Chengtu, China. After a day of rest, briefing, refueling, and getting organized, the crews would make their raid over Japan and return to Chengtu for the third stage. Then on the fourth day the bombers would return to India—four days for one mission.

Because ATC transport planes were flying day and night to supply the war needs of Generals Stilwell and Chennault, they were not able to ferry enough gasoline for the B-29s, a fact not realized in the original planning. Thus the Superfortresses were forced to ferry their own fuel. Seven flights were required to move enough fuel and supplies to the Chinese staging bases for one B-29 mission against Japan. This was expensive, time-consuming and risky for men and airplanes. For the first two raids, seven hundred thousand gallons of fuel were flown across the Hump.[21] Operations in China were always plagued with fuel shortages and weather reconnaissance had a low priority for fuel. A personal experience in 1945 proved this point: On one recon flight over the Hump, unexpected headwinds and weather slowed the flight, and we were forced to land at a British air base in Burma for refueling. The British, our own ally, refused to service the plane,—from American-made gasoline trucks we were denied fuel. All our persuasion, mutual friendship, lend lease, and storming around, and reminding the British that it was our gasoline in the first place, got us nowhere. We had to call our home base in India and have fuel flown in.

Charged with the responsibility of weather reconnaissance for the B-29s, a detachment of 2nd WRS B-25Ds was located at Sian (Siking), China, 300 miles from Chengtu. Although the B-25s could not fly as far nor as high as the huge bombers, they would take off before a B-29 raid to locate fronts or weather problems and climb through the overcast to determine icing conditions. For the B-29s, heavily loaded with gasoline and bombs, icing was extremely serious since the *Superfortresses* were not equipped with de-icing boots, or anti-icing fluid for the propellers.[22] Weather recon B-25s did have de-icing equipment and were able to determine icing conditions more safely than the B-29s. If the big bombers became heavily loaded with ice, their only alternative was to abort the mission, drop below the clouds, and fly around until the ice melted and return to Chengtu.[23]

Since the 2nd WRS had to fly its own supplies across the Hump, Colonel Ellsworth decided to stockpile three thousand gallons of high octane gasoline at Tiwah (Urumchi), the most remote Tenth Weather station in China. Tiwah was 185 miles southwest of Outer Mongolia, nestled in the foothills of the Great Shan Mountains, with peaks to eighteen thousand feet.[24] If the 2nd had enough fuel, synoptic reconnaissance between Tiwah, Chengtu, and Sian (Hsian) would be able to detect weather forming before it began its eastward drift, important in forecasting for the B-29 raids. Also the Tenth wanted and had repeatedly requested permission to locate a DR station at Tiwah, in an area controlled by the Chinese communists and near Japanese lines. The Chinese had refused the request.

To try to solve the fuel problem in China, Colonel Ellsworth, in Octo-

Maj. Harry Albaugh and Maj. Joe Dillow flew a C-109 tanker over the towering mountains of the Hindu Kush in a project to ferry gasoline to weather squadrons in China. *Photograph from collection of the author*

ber 1944, requested a C-109 tanker aircraft to ferry fuel to the remote stations. Another purpose of this mission was to try to find a "great circle" route from India—looking for "the back door to China."[25] This first mission to ferry gasoline turned into a "Chinese circus," and proves the truth of the warning Rudyard Kipling made when he wrote the epitaph, "A fool lies here who tried to hustle the East."[26]

With the necessary diplomatic permissions, Maj. Harry "Tex" Albaugh, pilot from Mariposa, California, with Maj. Joe Dillow, "copilot"/radio operator, from Oregon; and Lt. William Breeze, Ocoee, Florida, as navigator, took off from Barrackpore on 3 November 1944. Dillow was an electrical engineer and such a skilled radio technician that he was taken along because of the importance of communication in that remote region. Major Albaugh said, "Dillow could build a radio with 'three paper clips and some wire' and I wanted him on that mission."[27] After refueling at Agra, of Taj Mahal fame, they landed at Peshawar on India's (now Pakistan's) northwest frontier. At Peshawar, the aircraft tanks were topped with ninety-octane gasoline, the highest available.[28] As they flew over the mountains of the Hindu Kush and onto the Takla Makan desert, the C-109 had a load of 4,900 gallons of gasoline. Maps were marked "unexplored territory," and the pilots' efforts to find a passage through the mountains turned into blind alleys too high for the C-109 with its load. Precious daylight hours were spent looking for passes. The crew was often flying below some of the mountain peaks—towering above their flight path—seemingly higher than Mt. Everest. Since sea level pressure readings in the mountains were unknown, the plane's altimeters could not be set to indicate an actual altitude. This problem is dangerous when flying over unexplored areas with no weather stations to report local conditions. Many aircraft have flown into the mountains because their true altitude in instrument weather was not known.

With darkness coming on, Major Albaugh, at the end of a mountain pass, found a small town and flew low to investigate. Seeing a crude air strip and a wind "T,"[29] he landed with full flaps and stopped the huge plane just short of a stockade. Crowds gathered and, with deep bows, smiles, and sign language, the airmen were greeted by a Chinese general and his staff. Tea was formally served although none of the Chinese spoke English. With the help of a map, Major Albaugh found that they had landed at the village of Yenchi, Sinkiang Province, China. He sent Major Dillow to the plane to send a message to the British in Peshawar, asking that their location be relayed to Calcutta weather headquarters. The message was never relayed although the British, through diplomatic channels, asked what their Allies were doing so far north. At Tenth Weather headquarters, Colonel Ellsworth was angry at the British fumbling since, had the weather crew been thrown in jail, they might never have been found.

Far from being jailed, the airmen were honored guests. After a night of feasting , toasting with rice wine, and a local Chinese play, the tired, wine-soaked guests had a good night's sleep. Next morning an interpreter arrived, having traveled all night by horse—a distance of one hundred miles. The crew learned that their host was the commanding general of the Chinese 29th Army. Using the general's influence, the airmen asked that the field be prepared for the takeoff to Tiwah. Built three years before, it had never had a plane land. The runway was in poor condition and was only two thousand feet long. After the Chinese had cut the sage, filled the ruts, and tamped the earth, the crew lightened the plane's load by removing eight hundred gallons of precious gasoline. Still the odds were great that the plane could not get into the air in two thousand feet.

Next day, when the C-109 was ready for takeoff, the Chinese general, his chief of staff and three other members of his party, decided they would like to fly on the trip to Tiwah. Major Albaugh explained that they had tried to lighten the plane, and even without the five Chinese, it was possible the plane would not get off the ground and could crash. The general smiled and, pointing to Albaugh's crew, said they were brave men and he was willing to take the chance. After racing the engines, holding the brakes and with full flaps, the plane wobbled into the air. At Tiwah, the group was met by the American Consul Horace Smith and a Russian delegation. Mr. Smith had found some Russian gasoline for the flight to Chengtu and had been successful in getting permission for the DR station.

Yes, they had gasoline, the Russians explained, but because they were not at war with the Japanese, they could not give it to the United States. With no authority, Major Albaugh threatened to cut off American lend-lease unless they were sold gasoline. The Russians continued to say "no!" Albaugh even offered them the eight hundred gallons left at Yenchi. After some debate, they decided to lend the fuel to the Chinese who would lend it to the Americans. Two thousand gallons of gasoline filled the tanks of the C-109. However, the gasoline was only 70-octane, and the Chinese dropped blue pills into the tank saying it would increase the octane. With a full load of Russian gasoline, the flight crew, and the Chinese officials, the takeoff was successful and the aircraft engines did not cough or sputter. Albaugh remembers the incident of fifty-two years ago and explains that the C-109 had a small reserve tank of 100-octane gasoline which was used for takeoff and, at cruising altitude and speed, they were able to switch to the 70-octane.[30] After losing their way several times—maps were almost useless in upper China—the crew found the Sian beam and completed the trip to Chengtu. Later, on a return trip, the DR station was flown to Tiwah on the Tenth Weather C-47, "Betsy," and vital coverage was added to the Chengtu weather maps for the B-29 missions.[31]

This story seems strange and fictitious, yet my experiences in flying in extreme northwest China—actually in any part of China—made it believable. Flying Tenth Weather Squadron weather missions, I made several trips to Sian and Lanchow to supply weather stations in this starkly beautiful, unbelievably primitive part of the world. The people were more Mongolian than Chinese and were always smiling and happy. They wanted to please and become a part of the new world, yet their culture was so deeply flavored by the grace and charm of old China that their efforts were often comical. The Chinese always smiled and their smile and laughter were an enigma. They smiled when they were sad; they laughed when they were happy; facing death they smiled. As they watched an execution, they smiled and laughed and turned it into a social event. The Americans could never understand this.

One of the problems of weather reconnaissance was gaining the confidence of military commanders in an area where supplies were so pitifully short. Some commanders never thought that weather crews were making a real contribution to the war when they were "only flying weather." So, crews were constantly asked to do other "real" missions, especially rescue. A B-29 went down south of Shehsein, in Shansi Province, and the airmen were picked up by the Chinese communists. They were royally feted, and the communists even sent word to the Twentieth Bomber Command on the crews well-being and location. With the blessings of the communists, Capt. John Kuntz of the 2nd WRS's Flight C in Hsingching, flew a B-25 into a small airstrip in Japanese-infested territory and rescued the bomber's crew on 6 February 1945. Kuntz rescued another nine airmen on the same strip one month later. However, on 19 June 1945, on a mission for the Office of Strategic Services (OSS), Kuntz and his crew crashed into a mountain and all were killed.[32] There were political problems as well. Second WRS recon pilot Charles H. Broman, Tubac, Arizona, met Mao Tse-tung and Chou En-lai on one mission. As a result, he and others of the 2nd WRS had a "price on their heads" from the Chinese Nationalists. It was difficult to tell friend from enemy.

The 2nd WRS lost three crews and six aircraft on weather missions in China.[33] In situations that "make your asshole twitch," to quote another pilot, when I was flying weather in China, we flew in and out of many airfields controlled by the communists. Always, on landing and take off, we faced guns on the end of the runway. They could have shot us down, but they did not. Weather recon was a tactical and diplomatic minefield.[34] Because weather recon bases were located near Japanese-held territory and within Chinese communist control, protection was critical. Col. Jack Chennault (son of the famous commander of the "Flying Tigers"), his P-51 fighters, and a flight of four B-25s were sent to Sian to prevent the Japanese from moving too near weather recon airfields. All four of the B-25s were lost in fighting the Japanese.[35]

In China, the B-29 *Superfortresses* were big only in the minds of Washington military leaders and theater commanders. The B-29 was a great aircraft in the wrong theater. The problem of supplying fuel and armament to the B-29 bases in China was more than anyone had anticipated. Unimaginable distances over China and bitter weather were problems no other theater had presented. B-29s had not been designed for the extremes of weather and high altitudes required for flights over the Hump. The crews flew heroically, and many died unnecessary deaths. General Arnold's dream of a super long-range bomber in the CBI theater was turning into a military nightmare. Gen. Curtis LeMay and his *Superfortresses* were finding it impossible to fight an effective war against the Japanese, and in January 1945, LeMay moved his *Superfortresses* to the Mariana Islands, where fuel and ammunition could be supplied by ship.[36]

In the Marianas, when the B-29s were free from the supply problem and the icy, high altitudes over the Hump, plans could be made for bombing Japan. Stanford G. Kahn, meteorologist from Wilmette, Illinois, flew B-29s raids against Japan. He recalls, "Target weather was a problem, and weather reconnaissance was necessary to augment sketchy observations from China and the Soviet Union." When he moved to the Mariana Islands, General LeMay decided that weather reconnaissance was so important that he assigned fifteen B-29s from bombing missions to pure weather reconnaissance. While LeMay had faith in weather squadrons, he always wanted weather reconnaissance within his own command.

Kahn explained, "Thirty weather observers were assigned to the Twentieth Air Force, fifteen with weather experience and fifteen with special radar training. Five B-29s were based on each of the three islands, and three daily round-trip flights were made to Japan from June 1945 until the war ended in August. All flights made synoptic and route observations used to forecast target weather. Since the flights were over enemy territory, weather planes were armed, and a normal load of bombs was carried. However, on weather flights, the bombardier was replaced by a weather observer. Observations were made every half hour, and the information was sent after leaving Japanese territory." Kahn flew thirteen weather missions from Tinian during his assignment.[37]

General LeMay's chief weather commander was Lt. Col. Nicholas H. Chavasse with his 655th Bombardment Squadron, flying converted B-24s. Contrary to its name, the 655th was never a bombardment squadron, and from initial activation, it was equipped and crews were trained as a heavy weather reconnaissance squadron. Chavasse's 655th consisted of three flights of four B-24s each, including aircrews, weather officers, and support personnel. Each flight operated as an independent unit. The primary mission of the 655th was to scout weather conditions over B-29 targets and to support P-51s of the VI Fighter Command, based on Iwo Jima. The B-24s flew night missions and sent the information back in time for early morning briefings of B-29 crews. In addition there were two flights each day averaging twelve hours each, for synoptic data for weather forecasts.[38]

Howard Lysaker, Mascoutah, Illinois, weather observer/gunner in the 655th described Chavasse's B-24s: "The planes carried a ten-man crew and, except for the pilot, copilot, navigator and flight engineer, each crewman was a qualified gunner and manned a gun turret. Each of three bomb bays carried extra fuel tanks for long range missions. In the fourth bomb bay was the AN/APQ-13 radar console and operator. In addition to the weather radar, the aircraft carried a radio altimeter, a second pressure altimeter, airborne psychrometer, and a cloud-height meter. On one mission in the summer of 1945, while flying with Col. Chavasse, Lysaker scouted a typhoon threatening the islands. On a low level mission, the ocean surface was a foamy spray, and the aircraft was tossed about "like a ping pong ball" with altitude changes of sixteen hundred feet and sixty-mile-an hour airspeed variations. His crew was in the storm for three hours.[39]

In early days of weather reconnaissance, Colonel Chavasse and his 655th (later the 55th WRS) introduced the B-24 to weather recon and prowled the Western Pacific in 1945 as the original "*Typhoon Hunters*," flying out of Guam, Iwo Jima and Okinawa. The big Consolidated B-24 high-wing, twin-tailed "*Liberator*" matched the B-17 in size, speed, and range with four Pratt & Whitney R-1830-43, 14-cylinder Twin Wasp, 1200-HP engines with turbo superchargers. More B-24s in different versions, were produced than any other warplane. A total of 18,188 planes were built in World War II, at a cost of $215,516 each in 1944.[40] The B-24 was modified for other war-time duty: the C-87 for cargo and the C-109 as a fuel tanker. Consolidated had used a new aerodynamic design—the Davis wing, extremely long and thin, giving increased lift, maneuverability, and range.[41] The 655th weather pilots flew their huge planes like fighters whether they were skimming the water or flying at twenty-five thousand feet.

The B-24 had a long history in weather reconnaissance. Chavasse and his crews started training at Will Rogers Field, Oklahoma in late 1944. Long range, low-level flights from Will Rogers over the Gulf of Mexico and the Caribbean were part of their training. While not actively engaged in hurricane recon on training missions, they were under orders to observe and report cyclonic activity, tropical depressions, easterly waves, etc.[42] In 1943 and '44, even earlier than Chavasse's flights, the 17th Weather Squadron, in the South Pacific, furnished weather observers to fly B-24 bombing missions out of Munda Point, on New Georgia, Bouganville and New Ireland Islands. These were combat missions—weather was secondary. The 17th's B-24s were worn, overworked and in poor condition. Robert Lashbrook, weather officer from Ojai, California, remembers, "It was not unusual to go to two, or even three planes before the pilot could find one where he could get all four engines running. They could not get parts to keep the planes in the air."[43] On night missions, "I always looked for St. Elmo's fire, a spectacular, eerie, unearthly glow coming from the sharp edge of the plane's wing. Almost every pilot who has flown in any type of weather has experienced this fiery, corona discharge. Saint Elmo (Sant' Ermo) was the beloved Bishop of Formiae, and sailors who first observed this fire believed that it was a visible sign of the saint's guardianship."[44]

Even on bombing missions, crews made new discoveries, not related to war, that benefited all nations. As the B-29 *Superfortresses* flew their high altitude "rain of death" over Japan, they found unexpected and tremendous wind velocities. Bombardiers found wind speeds of 200 knots or more, and correcting for drift was almost impossible. Kent Zimmerman, San Antonio, Texas, B-29 pilot, found that the "big winds" would blow the planes off course forty to fifty miles in one hundred miles of flight.[45] Bombing runs had to be made directly up or down wind—cross-wind bombing was impossible. Bombing Japan's best defended cities in the face of a 200-knot headwind was suicide since the aircraft were almost motionless over the target. When bombing downwind, ground speeds in excess of 500 MPH made accuracy impossible.[46]

What was this "big wind"? Although they did not know it at the time, the B-29 crews had discovered a high-altitude, high-velocity stream of air, later called the "sub-polar jet stream." Guy Murchie, in *Song of the Sky,* describes this new discovery, ". . . navigators of B-29 bombers flying to southern Japan near the end of World War II . . . were amazed to find that, near the horse latitudes at 35,000 feet, they regularly met west winds of more than 250 MPH, three times as swift as the average hurricane. Sometimes they actually found themselves flying backward . . . in 400 MPH headwinds! Something big was up. An entirely new kind of wind had been discovered, . . . a super-charged "Gulf Stream of the upper sky"—invisible upper winds that crack the secret whips of weather command massive disturbances that never come down to earth."[47] In 1945, meteorologists did not know what the "jet stream" was, but later study found it to be a "major wind of the world," a dominant factor in all weather that passes over temperate earth.[48]

The Americans and Germans had stumbled onto this phenomenon at about the same time. A Nazi *Junkers* photo plane was flying at an altitude of 17 km over the eastern Mediterranean. The weather observer on the flight reported a 170-knot wind. Later a "Mayday!" distress signal radioed, "Strong headwinds—can't make it—have to ditch. . . ." The jet stream had claimed its victim.[49] American pilots described the winds as" though the air had been blown out of a gigantic nozzle," and the name "jet stream" seemed apt. Later the winds were described more like rivers of air, very wide and shallow, meandering about, accelerating and decelerating.[50] It was later learned that there are two jet streams, one above and one below the equator. In the northern temperate zone, the jet stream "sizzles through the skies over China, the United States and the Mediterranean area, weaving about in conformity with the changing pressure contours of the great waves of the northern polar front . . . with a rhythmic motion which sometimes follows the phases of the moon . . . "[51] This giant jet of air has been used by trans-oceanic aircraft, and continues to be used as a huge tailwind flying east and a wind to be avoided when flying west.

With two atomic bombs dropped on 6 and 9 August 1945, the war ended. The 655th had been redesignated as the 55th Reconnaissance Squadron, Long Range, Weather and remained in the Pacific to battle the typhoons. The famous CBI Hump was closed to freight traffic, and weather recon was no longer necessary. But stopping a war is not easy. We realized that we couldn't just automatically go home and leave all weather stations and personnel stranded. So, on 6 November 1945, flight personnel of the Tenth Weather Squadron were transferred from India to Shanghai, China. We made our last trip over the Hump, saying good-bye to India, tipping our wings to Kunming, China, and flying on to Shanghai, with no weather to worry about. From November until March 1946, we had some of the toughest flying we had encountered, flying heavily loaded C-47s, taking off from remote mountain bases in China loaded with twenty to twenty-five weather men, their personal gear, and instruments from their stations. Takeoff was always at dawn to get the best lift from the cold heavy air. We covered the whole of ancient and enigmatic China. We were too young and too anxious to get home to appreciate the priceless adventures we were having—from Lanchow to Sian, to the north of Kunming and to Canton. These last missions were uneasy, heavy with foreboding. To die in combat is a tragedy, but to die after the war, trying to clean up the mess of battle and marking time until you can go home, is a story of greater anguish.

One dream-almost-nightmare weather mission came on 28 November when, by order of Lt. Col. Arthur McCartan, 1st Lt. Fred J. Huber and I were selected to fly Tenth Weather Region officials from Shanghai to Tokyo to inspect Japanese weather facilities. Huber and I had flown hurricane reconnaissance in the Atlantic in 1944. On this Tokyo mission were S/Sgt. John Terrell, Tulsa, Oklahoma, and Sgt. Donald B. Carlson, Palatine, Illinois, flight engineers; and Sgt. Alex Novak, New York City, radio operator. From Shanghai, in Tenth Weather's C-47 "Betsy," we took off across the East China Sea, climbed over the clouds, headed to Tokyo, homeland of our enemy of only a few weeks before. We were anxious to see Nagasaki, which had been devastated by the second atomic bomb. However, clouds covered the city. Then, as the clouds began to clear in the distance, we could see the smoking ruins of Hiroshima—tiny plumes of smoke still rising from the blackened city, 112 days after the fateful bomb. We circled the city at one thousand feet for about ten minutes. I don't remember a word anyone said, but smoking Hiroshima spoke its message. No one ever forgets a scene where thousands had died to stop a war.

After the flight to Tokyo, the colonel in charge of the weather mission decided that we should to go to Keijo (Seoul), Korea, to "check on weather stations"—translated: "to see the geisha girls." So, on 1 December, at 1130, Huber and I headed the C-47 toward Korea, across the Sea of Japan—a five-hour flight with good winds. The weather forecast was "possible snow storms in Keijo"—that would be no problem, we were weather-seasoned pilots. By mid-afternoon, we were getting gloomy forecasts of heavy snow in Keijo although the airfield was still open. Our radio compass (ADF) was oscillating ominously, bracketing our heading. As the overcast darkened, we dropped lower and lower to stay in visual contact with the sea and started looking for the Korean coast—dark choppy water was everywhere. Then the message came that Keijo was closed with a snow storm. and we were denied clearance—I had to listen several times to accept the message. We didn't have enough fuel to return to Tokyo or Shanghai, and we were flying at about one thousand feet to stay under the clouds. Darkness was settling over. One rule of flying had been broken: Never put two first pilots in one airplane—either both will be in charge or neither will be. Fred and I were having problems trying to decide what to do. Fuel was low, clouds were dropping, dark was coming, and all we could see was black icy water with its siren call to death. Just then, we crossed the coast. Korea? We didn't know. Fate was kind. We found a small fighter airstrip on the coast. We cut the engines, lowered the wheels and made the best short-field landing any pilot had ever made, stopping at the last foot of runway. We were on the ground, and heavy snow was moving in on the field.

We didn't know where we were except that we were somewhere on the coast of Korea—at least, we hoped it was Korea. If natives around this tiny airstrip came to pick us up, would they know the war was over? Would they be friendly? Would we be taken prisoners of war after the war was over? It was bitchy cold outside and, with engines off, the plane had no heat. We sat in the plane and waited. In about an hour, a huge U.S. Army 6x6 came rumbling down a small road. We cheered the Army! We had landed at a small airstrip just south of Fusan (Pusan). Quickly calculated, we had flown against 50 to 60 MPH headwinds from the northwest and were two hours overdue at Keijo, still 150 miles to the north.

After an Army welcome, a hot meal, a warm night's sleep in an infantry barracks in Fusan, the quartermaster filled our tanks with 90-octane gas. We forgot about Keijo and the geisha girls and at 1215, 2 December, we headed for Shanghai, back across the East China Sea, landing five and one half hours later. This touch down was our final "weather recon" flight.[52]

The Second Weather Reconnaissance Squadron was deactivated at the end of the war when the Hump was closed. Yet, war or no war, weather continued in all parts of the world and the story of air weather reconnaissance continues.

4

The Weather Net—Equator to the North Pole
1948-1950

World War II ended quickly. On one special mission, on a fateful 6 August 1945, the B-29 *Enola Gay* dropped an atomic bomb on Hiroshima, Japan. Three days later, B-29 *Bock's Car* dropped a second, larger atomic bomb on Nagasaki. Two cities lay in ruins and over two hundred thousand people died. On 15 August, President Harry Truman issued "General Order No. 1" to Gen. Douglas MacArthur: "The Imperial General Headquarters by direction of the Emperor . . . hereby orders all of its commanders in Japan and abroad to . . . surrender unconditionally."[1] The Allies became the first to use air power to end a major war without invasion of the enemy's homeland.[2]

A brief history: As early as 1942, the Manhattan Project, under the direction of Brig. Gen. Leslie R. Groves, had been charged with bringing an atomic bomb into being. General Groves indicated that the bomb would be available during the summer of 1945. War planners, in September 1944, had ordered the 509th Bomb Group, with specially modified B-29s, to deliver the bomb. Based on intelligence that Germany would surrender without the use of the bomb, the United States selected Japan as the target. The 509th was deployed to Tinian, in the Mariana Islands, under the command of Gen. Curtis E. LeMay, 21st Bomber Command, Guam. Col. Nicholas Chavasse's 655th Weather Reconnaissance Squadron flew weather recon for LeMay's B-29 Superfortresses.[3] Before each of the two bombings of Japan, a weather reconnaissance plane took off one hour before the bombers to determine weather conditions over the target. Skies

The bombing of Nagasaki, Japan, brought an end to World War II. Floyd Lasley, 55th Weather Reconnaissance Squadron, took this photograph of a "colorless gray" bombed-out city on 20 September 1945. A large hospital ship is anchored in the harbor. *Photograph courtesy of Floyd Lasley*

were clear over Hiroshima. On the 9 August raid, the primary target was Kokura, with Nagasaki as alternate. Kokura was covered with clouds, so the bomber flew to Nagasaki and dropped its cargo. Six days later Japan surrendered.

The atomic age began on 16 July 1945 with a fiery blast when the world's first atomic bomb was exploded at Alamogordo, New Mexico. The sound was heard fifty miles away, and light was seen for two hundred miles. With this awesome "big bang," the new age of atomic power was born. On the next day, President Harry Truman, Prime Minister Sir Winston Churchill and Premier Marshal Joseph Stalin met at Potsdam, Germany, to plan the end of the war. The Japanese received the ultimatum to surrender on 26 July 1945. The "rest of the story" is well known.

As the world anticipated the end of the war, a life-and-death drama was being played out in the Pacific. Two P-51 pilots of *"Possum Red"* flight of the 460th Fighter Squadron from Ie Shima, were lost on a "cat cover"— long missions where fighters were sent to protect Air-Sea Rescue *Mariner* PBM-5s. Unable to locate themselves, the two pilots, Capt. George Elves, Fort Myers, Florida, and Lt. Tim Frable, Weatherly, Pennsylvania, flew until they ran out of gas. Eight hours into the mission, first Elves, then Frable bailed out. The two men bobbed around in life rafts for five days, fighting freezing nights, scorching days and sharks, with no food and no water after the second day. A C-54 passed overhead each day but did not to see the pilots' signals. Thirst was overpowering; exhaustion, lack of sleep, heat and exertion were taking their toll. On the fifth day, two B-24s, flying low, passed them by. Wait! About four miles away, the two big planes "stood on their wings" and turned back to the two rafts. The planes dropped water, radioed for help and stayed with the men until a Navy *Mariner* came in.

When the two men were safely in the rescue plane, Navy pilot, Lt. John Donahou, Winona, Illinois, asked, "Say, do you fellows know what happened yesterday?" "Hell no," Frable said, "We've been sitting out there on the Pacific for five days." "Good news," the cheerful pilot said, "The war ended yesterday!" [4] It was forty-five years later that Tim Frable learned that the B-24s were from the 655th Weather Reconnaissance Squadron, stationed on Okinawa, flying weather recon for the B-29s bombing Japan. Pilot of the lead plane was Col. Nick Chavasse, of Temple Hills, Maryland, and pilot of the second plane was Capt. Frank O'Leary, Bloomington, Delaware. Radarman Carlos Arrobio, Glendale, California, had found the men as a radar blip, and radio operator Tom Letz, Big River, California, had seen the flash from the mirrors. Later, at a reunion of the crews, Colonel Chavasse remembered his thought when they found the men: "after a year of flying fifteen-hour missions over miles of open ocean, gathering weather information, that mission made our war."

The last mission of the war may have been flown by a weather reconnaissance B-24 of the 20th Air Force. Piloted by 1st Lt. Joseph L. McCarter, Wadley, Alabama, the plane took off from Iwo Jima at 2145, 14 August 1945, fifteen minutes before the news of the end of the war was received at the base. After takeoff, radio operator T/Sgt. Clarence R. Williams, Bishop, California, turned to *The Voice of America*, from Chunking, China, and heard the news. He tapped navigator, 1st Lt. William O. Brown, Jr., Macon, Georgia, on the shoulder and, with suppressed enthusiasm, said, "Beg pardon, sir, the war is over," and pointed to the radio.

"Navigator to crew! Navigator to crew! The war is over!" Brown yelled into the intercom. There was stunned silence. Tail gunner S/Sgt. Silvio J. Former, New Brunswick, New Jersey, broke the silence: "What the hell are we doing here? Let's go home." "It may be some kind of a trick," S/Sgt. William F. Zoulek, Traverse City, Michigan, answered. The crew continued its weather mission over Tokyo. (Broken overcast and clear over Nagoya.) Circling over Nagoya, the Japanese lit up an airfield for them. Copilot 1st Lt. Gerald K. Ossinoff, New York City, was cautious, "Maybe they don't know about it and are sending up fighters." "Heck naw, they think we are Eleanor," said S/Sgt. James K. Suttles, Lumpkin, Georgia. "I wish I could speak Japanese," said weather officer 1st. Lt. Robert P. Sheuring, Greenwood, Wisconsin, "I'd call down and ask for a complete weather report." The lights of the field flicked off, and the B-24 weather plane headed back to Iwo.[5]

As wars end there are always strange and interesting stories of missions that should not have been flown, last missions, and extravagant flights of celebration. One such mission was remembered by Lt. Col. John Hug, B-29 aircraft commander from Boise, Idaho. On 2 September 1945, for the surrender ceremony aboard the battleship USS *Missouri*, the Air Force ordered every available B-29 in the air over Tokyo. Accompanied by the roar of six hundred B-29s circling overhead, General MacArthur made a two-minute address, heard over the *Voice of America*. "It is for us, both victors and vanquished, to rise to that higher dignity which alone benefits the sacred purpose we are about to serve. . . . A better world shall emerge out of the blood and carnage of the past."

"My crew and I were over Tokyo for six hours," Lieutenant Colonel Hug remembers. "You can imagine six hundred B-29s over Tokyo, going in every direction. My eyes bugged out dodging and looking for airplanes—every aircraft was on its own. . . . You can imagine what a heyday that was. When we landed, we had flown an eighteen-hour mission." Hug's regret was that he did not fly low over Hiroshima to see the smoking ruins—the symbol of the war's end. "I asked the crew if we should go down and look at Hiroshima. Everybody was so damned dog tired, the majority voted 'no,' and now I still wish we had gone to see it."[6]

Weather reconnaissance had proved its value in wartime and, with the coming of peace, the First Air Weather Group (Provisional) was organized 13 July 1946 to administer three new weather reconnaissance squadrons (WRS). The 54th, 55th and 59th WRSs were assigned to cover weather trouble spots of the world. The 53rd WRS was already operating in the Atlantic and Pacific; the original 654th Bombardment Squadron in England, was designated the 54th on 4 September 1945 and sent to Guam. The war-time 20th Air

Boeing WB-29 *Superfortress* from the 55th Weather Reconnaissance Squadron. The WB-29 brought a new dimension to weather recon and served in all parts of the world in every kind of weather for thirteen years. *U.S. Air Force photograph courtesy of John W. Pavone*

Force 655th Bombardment Squadron was redesignated the 55th WRS, Long Range, on 16 June 1945 and also stationed at Guam; and the 59th WRS, Very Long Range, was activated on 10 August 1945 at Will Rogers Field, Oklahoma.[7] The lineage of these squadrons is a confusing tangle of activations, de-activations, "on paper" designations, redesignations and name changes. Historians grieve at the lack of agreement in the various "official" histories.[8] (See Appendix)

In peace, as new Air Force missions required special weather forecasts, the responsibilities of air weather recon were broadened, and many new assignments were given to the weather crews thinly scattered around the world. Also, weather recon got better aircraft. Except for the B-29 flown by General LeMay's 20th AF, the North American B-25D, Boeing B-17, and Consolidated B-24 were the aircraft being used in weather reconnaissance. Immediately after the war, battle-retired B-29s became available, and many squadrons were pleased to began using converted RB-29

Superfortresses, capable of ten- to twenty-hour missions. As weather history is written, each new aircraft added to the weather fleet, was a historical milestone and opened new operational capabilities and opportunities.

The RB-29 went to work immediately and flew its first weather recon mission on 25 May 1946 with the 59th WRS from Castle AFB, California. Although used only fourteen months in battle, this super airplane, in its new role, brought in the "golden age" of weather reconnaissance. The RB-29 flew higher, faster and farther than any other recon aircraft. The *Superfortress*, the largest of the World War II bombers, has been called "the greatest U.S. gamble of the war." Development and production of the B-29 cost $3 billion against $2 billion for the atom bomb, and the plane was ordered right off the drawing board, without a prototype. Time was critical.[9] From drawing board to the sky took two years, agonizingly slow in war time, but incredibly fast for production of such an advanced plane. At one time in 1944, eight *Superfortresses* were rolling off the assembly line each day. A total of 3,760 planes were built at a cost in 1944 of $605,360 each.[10]

The Boeing B-29 was the Air Force's largest aircraft, 99 feet long, with a wingspan of 141 feet, 3 inches and a height of 27 feet 9 inches. Four 2,200-HP 18-cylinder Wright Cyclone engines, the most powerful ever designed, pulled the seventy-ton plane and eleven crew members through the sky at 300 MPH with a ceiling of thirty-five thousand feet and a range of five thousand miles. To reach this performance, engineers faced an aerodynamic puzzle of propelling a mass more than twice the weight of the B-17 at a greater speed. As speed is increased, the power needed is the cube of the velocity. To double the speed, eight times more power is needed. Then, when the weight is double that of the B-17, more drag is created which demands even more power or a new design. To solve the problem, Boeing created both more power and a new aerodynamic design.

Engineers designed enormous wing flaps—larger than the wings of most fighter planes—the largest relationship of flap to wing area of any airplane at that time. In conventional design, flaps usually do not increase the wing area—but are lowered to increase the wing's angle of attack for more lift. The B-29's huge hovering flaps created more wing area to provide increased lift at crucial moments during take-off and landing. In flight, flaps would retract back into the wings to reduce drag.[11] B-29s were designed and built in utmost secrecy, beginning from Air Force commander Gen. H. H. Arnold's dream in 1940. On 15 June 1944, a news bulletin proclaimed to the world, "The Air Force's mightiest weapon—the B-29 *Superfortress*—struck a late evening blow against the Japanese, the first strike against Japan since 18 April 1942, when Jimmy Doolittle's carrier-based B-25s hit Tokyo."[12]

Well-suited to the long weather missions, the B-29s had pressurized crew compartments designed for comfort. In the cockpit—a huge flying solarium—aircraft commander (pilot) and copilot sat side-by side with only a simple set of flight instruments; behind them the flight engineer had a complex panel of engine and systems instruments and the navigator had his own navigation instruments. Each crew member had a separate work station. The weather observer sat down front between the two pilots in the bombardier's position. The sound-proofed cabin allowed crew members to talk with each other without intercom. Sharing hours of tense intimate work, flight crews developed a special relationship in the air and sent unconscious signals needing no words. For example, Eugene Wernette, aircraft commander of the 57th WRS, from Fort Worth, Texas, often referred to David Magilavy, weather observer from Newport Beach, California, as his "third pilot in charge of landings." Magilavy's position was in the nose, directly in front of the two pilots. Someone asked Wernette if the B-29 was difficult to land. He said, "It's no problem. I just let Magilavy tell me what to do. On final approach, I watch his body signals. If he raises his left shoulder, I raise the left wing; if he raises his right shoulder, I lift the right wing. And when he slumps in his seat, I know it's time to flare out and haul back on the yoke. Magilavy makes a good landing!"[13]

In addition to the "front office" crew, the radio operator and the radar operator were in the aft position in the B-29 and were in constant communication with the front crew. There were left and right scanners whose jobs were to watch the engines on their side of the B-29 and keep an eye over land or ocean for downed aircraft, landmarks, giving the aircraft commander two additional sets of eyes. Dropsonde operators had positions in the rear compartment to launch the weather dropsonde and record its messages. Often the dropsonde operator and scanner stations were the same.

An example of new post-war special assignments was providing weather recon support for an Air Force special project. In mid-1946, as military aircraft began flying longer missions, and with the perfection of in-flight fueling, the Air Force proposed an experimental ten thousand mile, non-stop, over-the-pole flight from Hawaii to Cairo, Egypt. This mission, called *Paucsan Dreamboat*, would be flown in a specially modified B-29, named *Dreamboat*, piloted by Col. C. S. Irvine. To make the record-breaking flight, *Dreamboat* would need a tail wind averaging ten knots, and weather planes were asked to find and recommend flight levels with the best winds. Weather recon for the *Dreamboat* project was coordinated by Col. Nick Chavasse, chief of the Air Weather Service's reconnaissance. Chavasse organized a fleet of weather planes, stationed along the overwater parts of the route, and after many delays, the *Dreamboat* flight was scheduled for 4-5 October 1946.

Weather crews, still flying B-17s, made daily observations out of Keflavik, Iceland, and Søndre-Stømfjord (BW-8), Greenland. One aircraft would depart BW-8, flying to coordinates over Baffin Island, then across the Greenland ice cap to Keflavik. A second weather B-17 would fly from Keflavik to BW-8 for the next day's mission. Lt. Edward Vercelli, Joliet, Illinois, navigator on the 53rd WRS B-17, wrote, "Our mission for *Dreamboat* was to get a sequence and weather pattern for that area over a period of months."[14] Meanwhile, on the other side of the Arctic Circle, in Alaska, aircraft of the 59th WRS were also making daily observations for *Dreamboat*. The two weather crews flew a combined ninety-five hours from Whitehorse, Canada, in support of the *Dreamboat* project.[15] On the last leg of the flight, as *Dreamboat* crossed the North Sea, on 5 October, Lt. Leonard Winstead, weather officer in a 53rd WRS B-17, in radio contact in flight, recommended that the pilot of *Dreamboat* change altitude to get the increased tail winds needed and the flight was completed to Egypt with fuel to spare.[16] The B-29 was finding its place in peace missions.

Ray Wagner, of the American Aviation Society, writes, "A few aircraft have made such an impact on history that names like *Mustang* and *Thunderbolt*, and numbers like B-17 and B-29 have become a conspicious part of life in the 20th century."[17] Although used as a weapon against only one nation in World War II, the B-29 *Superfortress* made its name on one mission. In peace, in thirteen years of weather flying, the *Superfortress*, as the WB-29, was used to help all nations and earned the respect of weather crews. Someone in the 59th WRS estimated that in 1946 the average RB-29 mission cost $7,000, and crews were conscious of that figure—it was posted in many briefing rooms.[18] During the B-29's weather service, there were eight major accidents with a loss of fifty-eight crewmen.[19]

Atomic power was born to become the slave or peril to its human masters. Following the war, in June 1946, the United States began a series of atomic tests that would determine the safe and orderly use of atomic energy. Weather reconnaissance had a new assignment. Twelve hundred miles northeast of Guam is Bikini Atoll, an isolated island of the Northwest Marshall Islands, covering about two and one half square miles. Bikini was to be the first site for atomic test explosions.[20] This operation was called *Crossroads*. For the success of the bomb tests, weather reconnaissance crews had important questions to answer: What are the most favorable weather conditions for the bomb drops? What is the drift pattern and direction of the radioactive cloud? What is the radioactive cloud's effect on weather after the drop? What is the fallout pattern down range from the explosion? These were important questions never before faced by weather observers.[21]

A flight of B29s from the 59th WRS, from Castle AFB, California, joined the 509th Composite Group at Kwajalein Atoll in March 1946. Their mission was to check the atmosphere for background radiation prior to the four atomic test blasts for comparison to post-blast radiation. Weather recon flights were conducted daily on one of several tracks. Each of the crews flew approximately thirty twelve- to fourteen-hour missions, and the flights provided synoptic data for forecasts for the blast site and down-wind atomic cloud drift.

Charles Monnasee, 59th weather observer, from Omaha, Nebraska, recalls, "Each of the crews got to see one of the blasts, and our crew witnessed an underwater test from the air only a few miles away. The eruption of water looked like a giant wedding cake coming out of the lagoon when the bomb went off."[22] This "tourist-like" observation may seem naïve, compared to the importance of the tests. However, weather crews could do little more than fly the aircraft and watch the explosions. Detection equipment was installed aboard the aircraft and everything was so classified that the crews had little knowledge of what they did or who got the data.[23]

A second series of atomic tests was conducted in 1948 under the name of Project *Sandstone* at Eniwetok. In this series, Maj. Paul Fackler accidentally flew through an atomic cloud,[24] in what is believed to be the first aircraft and crew to fly through atomic debris in the atmosphere. The success of the *Crossroads* and *Sandstone* blasts, and the threat of radioactive clouds drifting around the world, increased the importance of weather reconnaissance and added atmospheric surveillance to the missions of weather squadrons. Special filtering devices, called "bug catchers," were attached to all B-29s to collect radioactive material from the atmosphere for analysis.

A special type of atomic radiation in the atmosphere was accidentally detected on 3 September 1949. An Air Force weather reconnaissance plane, flying at eighteen thousand feet from Japan to Alaska, had detected signs of intense radioactivity over the North Pacific. Planes elsewhere over the Pacific had reported radioactivity as much as twenty times normal. After analysis, on 19 September, in Washington, David Lilienthal, Chairman of the Atomic Energy Commission under President Truman, received the information that atomic debris had been picked up by an aircraft in the Pacific. The next morning, Lilienthal reported to the President a "whole box of trouble." The radioactive cloud was tracked by the Air Force from the Pacific nearly to the British Isles, where it was picked up by the RAF.[25] Three days later, after all data had been analyzed and atomic experts, including J. Robert Oppenheimer, had concurred, President Truman announced, "We have evidence that . . . an atomic explosion occurred in the USSR. . . ."[26]

The weather crew that discovered the international secret on 3 September was on a routine mission. Here's how it happened: 1st Lt. Robert Johnson of the 375th WRS began a synoptic mission on 30 August, from Yokoto, Japan, to Eielson, Alaska, but had to abort because of engine failure. He took a standby aircraft on 31 August, but was forced to land at his alternate airport at Misawa and cancel the mission because of an inoperative Ground Controlled Approach (GCA). On 3 September, Johnson took B-29 44-62214 from Misawa for a thirteen-hour *Loon Charlie* mission to Eielson. When the filters from his flight were analyzed, they suggested that an atomic bomb had been detonated.[27] Ironically, fate had scrubbed two missions and the third was successful. A successful mission earlier might not have detected the slow-moving atomic cloud.

The crew for this historic mission, in addition to Lieutenant Johnson, included copilot 1st. Lt. Lawrence S. Paul, navigators 1st Lts. Charles E. Massey and Gene A. Culbertson, weather observer 1st. Lt. Robert L. Lulofs, engineer M/Sgt. James R. Boswood, radar observer M/Sgt. Thomas R. Richardson, radio operators S/Sgt. Steve T. Yapuncicle and Pfc. William M. Kelley and dropsonde operator Lauren G. Cackstetter. S/Sgt. Michael D. Boyce was the crew chief.[28]

Little known inside stories that would never make official records add flavor to history. The story of "Gooney 263" is told by Jack Shelton, maintenance officer from Dallas, Texas. Gooney 263 was a Douglas C-47 (the C-47 was nicknamed the *Gooneybird*), an aircraft of the 2078th Weather Reconnaissance Training and Special Projects Squadron. Gooney 263 was equipped with extremely sensitive electronic equipment for detecting radioactivity coming from stacks at atomic reactor facilities. A young man by the name of Gene Harland was a navigator/radio operator and skilled electronics technician who operated the detection equipment and spent much of his time flying around the country checking reactor stacks. The plane and crew had moved to Tinker AFB, Oklahoma, where Shelton had Gooney 263 stripped for a major overhaul. Even the wings had been removed. Shelton got a call from Lou Cole, operations officer, and Arthur McCartan, squadron commander, saying that 263 had to be ready to go immediately for urgent sampling missions. Somehow Jack and his crew got Gooney 263 put back together and flight tested by the next morning. It turned out that the urgency came from the 56th WRS out of Yokota, Japan, that one of their planes had picked up atomic debris in the atmosphere, possibly a Russian atomic bomb. The Air Force Office of Atomic Testing (AFOAT), coordinator of the sampling program, apparently needed some data peculiar to the equipment capability carried by Gooney 263.[29]

Why were weather crews concerned with this type of activity? On 15 September 1947, Chief of Staff General Eisenhower, in a secret assignment, had charged the Air Force with responsibility for operation of a world-wide atomic detection system: "To determine the time and place of all large explosions . . . anywhere in the world and . . . leave no question whether or not they were of nuclear origin." In turn, the Air Force assigned this highly classified operation to the Air Weather Service since AWS was already operating B-29s for weather observations and could easily handle the additional assignment of atomic detection.[30] The Air Force Office of Atomic Testing (AFOAT), a staff agency of USAF headquarters, coordinated detection efforts, with close cooperation between the Atomic Energy Commission (AEC) and the Air Weather Service. Later, an operating unit, the 1009th Special Weapons Squadron, was established in Washington. In an interview with John Fuller, historian for the American Meteorological Society, Col. Don Yates, explained, simply, why weather recon was selected: "We had the facilities to do it." Weather planes, with only minor variations in mission tracks, showed no indication that they were doing anything other than routine synoptic weather reconnaissance.

When the Army assigned the first B-29s to weather recon squadrons, commanders were still jittery from the war, and weather recon was assigned a secondary mission of bombardment and the first weather *Superfortresses* carried a full complement of armament. But, on 30 August 1950, the prefix "W" was authorized for aircraft modified for weather—the B/RB-29 became the WB-29—and armament was removed.[31] The WB-29 had a special stamina in fighting nature's weather violence. On 8 September 1950, Capt. Charles Cloniger, of the 514th WRS, after losing one engine on his WB-29, *The Typhoon Goon,* continued its mission into Typhoon *Wilma.* Realizing that the typhoon's strength was important to the U.S. Inchon landings in Korea, Captain Cloniger completed his storm mission on three engines. For this he was awarded the Distinguished Flying Cross—the first ever to be awarded a weather pilot for a storm mission.[32]

In the fall of 1951, the 57th WRS sent a flight of WB-29s to Kirtland AFB, New Mexico, to participate in low yield atom bomb tests at the Nevada test site. These tests were called *Buster-Jangle.* Cloud tracking began northward toward Salt Lake City, Utah, but the cloud was lost in a snow storm and the aircraft returned to Kirtland. Two days later, Eastman Kodak Company, Rochester, New York, reported fogging of photographic film related to a snow storm. The cloud had been carried by the jet stream over New York about thirty-six hours before the film fogging incident. This report caused a stir in Washington and led to the discontinuance of all above-ground testing in the United States.[33]

Some of the 57th WRS WB-29s dipped into atomic clouds when they flew reconnaissance for the Nevada atomic tests in early 1951. Aircraft commanders for these flights were Roy Ladd, Tommy Mull, "Red" Hilgeford, and Ira Watson. Two aircraft from the 55th WRS also participated in these tests, piloted by Pierce Bilyeu and Tony Garrison. After these tests, sampling was done underneath the cloud overhang unless the aircraft flew into the cloud accidentally. Officially there were no more penetrations in the main atomic clouds following the Nevada tests.[34]

As atmospheric monitoring became a major Air Force responsibility, Lester R. Ferriss, Jr., former commanding officer of the 59th/375th WRS, explained the organization of the atomic sampling reconnaissance flights. In 1952, the Atomic Energy Detection System (AEDS) was established and the deployment of the recon squadrons was reflected in a plan called "Priorities of Programmed Units." The 56th WRS, Japan, had the highest priority because it was operating in the area where interception of atomic debris coming out of the USSR was most likely. The 56th was also conducting reconnaissance for the Korean effort. The AEDS's first line of defense was provided by the 54th WRS, Guam; the 56th, Japan; and the 58th, and 59th, Alaska. The second line of defense was by the 57th WRS, Hawaii; and the third by the 55th, California. The 53rd WRS, Bermuda, had a detached flight operating out of Saudi Arabia toward Pakistan to cover that area.[35]

The 1009th Special Weapons Squadron was located in downtown Washington, with tight security and excellent communications to support its role in atmospheric mission planning and coordination. The planning staff was composed of experienced weather recon officers, including Colonel Ferriss, Jim Chiarella and Roy Connell from AWS headquarters, and Margie Garrison, former 55th WRS officer. Ferriss records their first intercept, ". . . the phone rang in the middle of the night and the word was, 'get down here now!' Speeding down the highway toward Washington, I was stopped by two motorcycle police. 'What's the hurry?' 'An emergency!' 'What kind of emergency?' 'A national emergency!' They released me and told me to pay my fine later. I did."[36]

Squadron 1009 had clout. At one time, a mechanical problem grounded all B-29s, wherever they happened to be. The 1009th persuaded the Air Staff to give all AWS WB-29s a waiver to return to their home bases where repairs could be expedited." Former Col. Lewis Howes, in his book, *The Nuclear Explosion Detection System*, wrote, "It is sobering to speculate the course of events had there been no monitoring system. . . . The Soviet success would have been unknown to us [and] we would not have made an early attempt to develop the hydrogen bomb [thus] enabling us to maintain military superiority."[37]

To cover the vast open-water areas of the globe, the WB-29 made possible the organiztion of a world-wide weather net of synoptic flights. Bet-

Headquarters for the 56th Weather Reconnaissance Squadron in Yokota, Japan. The 56th flew WB-29 weather missions over Korea in 1950-51 to support the South Korean armies. *Photograph courtesy of Charles R. Hoyle*

ter weather reports were needed than were available at that time. And, perhaps unknown at the time (perhaps not), would be the urgent need for flights to track the drift of a new atmospheric ingredient—radioactive clouds.

Weather squadrons, stationed around the world, adopted names of birds to identify their synoptic flights and special missions. *"Ptarmigan!"* *"Buzzard!"* *"Loon!"*—strange names! These special flights collected observations over areas impossible to cover otherwise. The following flight tracks, over areas of open water in the Atlantic and Pacific, discovered and scouted weather patterns and storm systems over the uncharted oceans before they began their slow migration toward populated areas.

"Gull" flights were flown from Bermuda by the 53rd WRS, covering the South Atlantic's spawning areas for tropical storms. Flights were scheduled on Monday and Friday each week during hurricane season. *"Gull Quebec"* went north from Bermuda, turned east to the 35th meridian, then

flew south, forming a rectangle, and returned. "*Gull Papa*" dipped south from Bermuda to the Windward Islands, southeast of Puerto Rico, then east to the deep South Atlantic and returned to home base. Routine synoptic flights were given second-priority when tropical cyclones formed and became a threat to the Caribbean and the U.S. coastal areas.

"*Falcon Alpha*" missions were flown by the 53rd out of Burtonwood, RAF Station, England, covering the North Atlantic between Iceland and Scotland to scout weather affecting Europe. "*Falcon Echo*" covered the South Atlantic off the coast of Spain and Africa toward the Azores. Hurricanes and weather pressure systems from the South Atlantic moved toward Europe from the Azores. "*Falcon Special*," the B-17 that supported the *Dreamboat* project was—as the name suggests—a special flight of "*Falcon*" missions out of England.

"*Vulture*" was the name of three tracks from Anderson AFB, Guam, covering areas where typhoons threatened Japan, the Philippines, China, and the Mariana Islands. "*Vulture Lima*," "*Vulture Tango*," and alternate "*Vulture Oscar*" covered thousands of miles of ocean in a clover-leaf pattern out of Guam. These missions were flown by the 54th WRS.

"*Petrel Alfa*" and "*Petrel Bravo*" were 2,600-mile, 14-hour flights that went to "nowhere." Flown by the 57th WRS from Hickam, Hawaii, the *Petrel* missions were two of the most important and two of the dullest flights in the Air Force—both consisting of long hours "over nothing but water, water and for variety, more water."[38] Routine weather reconnaissance had no romance in it,[39] but the *Petrel* mission was extremely important—"like having dozens of weather stations out in the ocean, recording weather data, impossible to obtain otherwise."[40]

"*Buzzard*" flights were made from Yokota AFB, Japan, by the 56th WRS, covering typhoon areas from Japan to Taiwan; "*Buzzard Hotel*" flew northeast and "*Buzzard-India*" flew southwest from Japan.

From McClellan AFB, California, as far west as Hickam, Hawaii, "*Lark*" flights were made by the 55th WRS, observing weather from the Pacific Northwest, where observations helped forecast the Pacific and Arctic cold fronts that move across Canada and the forty-eight states; "*Lark Tango*" was an eight-hour loop west from California, "*Lark Victor*" was an alternate mission to Hawaii to the southwest and "*Lark Uniform*" departed McClellan to the west, turned south, and continued southwest to Hawaii.

"*Stork*" was a track, also from McClellan, that paralleled the northwest coast of Canada and went into Ladd AFB, Alaska, flown by the 55th WRS as another early warning shield against cold, wet Pacific fronts moving into Canada and the U.S.

"*Loon Hotel*" was a flight of the 55th WRS, which originated in Hickam AFB, Hawaii, and ended in Ladd AFB, Alaska; and in the summer

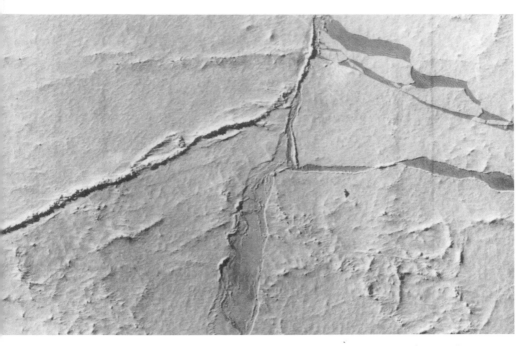

The barren desolate North Pole, photographed from a WB-29 on the 200th *Ptarmigan* recon mission. Temperature in winter often reaches below minus 90°F. There is no land—only frozen Arctic Ocean. *U.S. Air Force photograph courtesy of Lester Ferriss*

of 1948, "*Loon Baker*" missions were initiated. This round trip track was flown at eighteen thousand feet from Ladd Field, Alaska, to Nome, to Shema and return to Ladd.

The "*Ptarmigan*" flights, unique for the sheer drama of the savage weather wastes of the frozen Arctic, were the most significant and interesting of all synoptic weather missions. They penetrated uncharted areas and uncovered new information in an unknown part of the world. *Ptarmigan* was a grueling flight, from cold Alaska to the frigid North Pole, a region known as "the weather factory."[41] *Ptarmigan* flights, named after a far-ranging Arctic bird, began 17 March 1947 and were first flown by the 375th WRS. Starting from Ladd AFB, Fairbanks, Alaska, the huge RB-29s headed straight to the North Pole and returned to Ladd, a weary 3,760 miles and seventeen hours later.[42] The Arctic still remained a weather mystery, and little was known at that time about weather above 70° latitude. *Ptarmigan* flights provided basic research to lay the foundations for long-range forecasting for North America and helped unlock many secrets of the Arctic-born blasts that sweep across Canada and the United States.[43]

Lt. Col. Karl T. Rauk, La Crescent, New Mexico, was flight commander for the first *Ptarmigan* polar flight, with a crew of fourteen and one two-month old puppy named *"Yukon,"* the first dog to cross the North Pole since Rear Adm. Richard E. Byrd took his terrier *"Igloo"* across on 9 May 1926. The land, water, and icescapes on *Ptarmigan* were unsurpassed for their stark beauty. At ten thousand feet, one of the three navigators, Robert Sherr, Los Angeles, remembered, as he watched the sun disappear over the ice, a Robert Service poem, "I've watched the big husky sun wallow/In crimson and gold and grow dim."

Other crewmen on this historic first flight included pilot Maj. Roscoe B. Blackledge, Jet, Oklahoma; copilots, 1st. Lt. Clyde W. Courtney, National City, California and 1st. Lt. John Gagierd, Reading, Pennsylvania; navigators 1st. Lt. Donald M. Austin, Rochester, New York, and 2nd. Lt. Earl D. Depew, Detroit, Michigan; flight engineer, M/Sgt. Vernon C. Manuel, Mora, Minnesota; engineer-scanners, T/Sgt Vern O. Miller, Onalaska, Washington, and M/Sgt. Ennis A. Hilbrant, Phoenix, Arizona; radiomen, S/Sgt. James R. Spicer, Sidney, Iowa, and Cpl. Oliver M. Brown, Gorham, New York; and radar operator, S/Sgt. Harold L. King, Marion, Ill.[44]

Charles Augustus Lindbergh, first to fly solo non-stop across the Atlantic in 1927, was serving as an aviation consultant in the Pacific in the winter of 1947 and was a special guest of the 59th-375th WRS on a *Ptarmigan* polar flight. Lindbergh was anxious to experience the flight conditions encountered. The B-29 flight was from Shemya AFB, Alaska, over the Pole and into Yokota AFB, Japan. Crew members for this historical flight included 1st Lt. Jim Townsend, aircraft commander; Maj. Bob Foley, copilot; 1st Lt. Charles Markham, weather officer; 1st Lt. Dale Flowers, navigator; 1st Lt. Bob Branch, radar navigator; 2nd Lt. "Smitty" Smith, flight engineer; and M/Sgt. Dale Poteet, flight engineer. Lindbergh was listed as honorary pilot.[45] Most of the crew members were in their pre-teens when Lindbergh became a world hero, and, according to John Matt, author of *Crewdog,* a personal history of weather reconnaissance,"The whole idea, boggles the mind. To have Lindbergh in our midst, brief him about the squadron's polar navigation procedures, send him on a trip to show him how it is done."[46]

In the polar areas, weather observer Charles A. Monnasee, Omaha, Nebraska, and his crew fought the rigors of cold weather operations on *Ptarmigan* flights and found that the intense cold created new and unusual flight conditions. According to Monnasee,"We were the first RB-29s operating out of Alaska. The plane had been modified for arctic operations by the installation of Curtis Wright electric, fully reversible props to help

Flight engineer S/Sgt. Thomas K. Thrower presents a bucket of snow to 373rd Weather Recon Squadron commander Lt. Col. Clyde A. Ray after a nonstop flight from Alaska. Looking on is pilot Lt. James A. Ashcraft. *Photograph courtesy of Thomas K. Thrower*

brake the aircraft on icy, skidding runways. In the winter, heavy ice fog, caused by the prop blast on takeoff or landing would close the field for several minutes, and propellers kicking up an ice storm was a new experience."

The takeoff weight of a B-29 for a *Ptarmigan* flight was about 134 thousand pounds, including 7,800 gallons of gasoline, and two tons of survival gear. Weather flights pitted the sophisticated strength of technology against the brutal elements of nature. On one flight, Monnasee and his crew lost an engine shortly after reaching the North Pole. He couldn't remember the nature of the engine problem,[47] but John Hug, pilot on the *Ptarmigan* flight, did remember. Hug reports, "We had just turned south and the No. 3 engine caught fire. We feathered the prop, shut down the engine, but the fire kept going, with flames streaming back to the tail. You can imagine a fire being fed by a two-inch stream of gasoline, fanned by a three hundred mile-and-hour wind—an aerial blast furnace. The flight engineer quickly shut off the fuel manifold feeding gas to all the engines and reset the fuel feed for the other three engines from tank-to-engine. The fire went out and the mission was saved. We were ready to bail out—the nose wheel hatch was open. But I'm glad we didn't bail out—there was no place to go—we would still be walking. We ended up flying to Eielson AFB, Alaska, on three engines."[48]

To break the monotony of long polar flights, crews welcomed other assignments. John Gilchrist, Lutz, Florida, reported ". . . on spring and fall *Ptarmigan* flights, the Navy would send a photographer to take pictures of the Arctic Ocean so they would be able to forecast the spring ice thaw or the first winter freeze, to get ships in or out. We also tracked the 'ice island' as it drifted."[49] Lt. Col. Joseph O. Fletcher, commanding officer of the 375th WRS, speaking to an Alaskan Science Conference, in Washington, in November 1950, described the huge "ice islands" discovered on earlier weather recon flights. These monster chunks of floating ice, drifting a mile or so each day, were as large as eighteen miles across, thirty-five miles long and possibly 350 feet thick—large enough for weather stations or even aircraft landing strips. Colonel Fletcher showed a photograph of a four-engine Russian plane that had landed on the Arctic ice thirteen years before, indicating that the Russians had begun to study conditions along a short air route over the Polar cap long before the United States."[50]

"The polar regions are the only great gaps remaining in the world's weather network," Colonel Fletcher said. "Little is known of temperature and wind conditions at sea level in the Arctic. Although many airplanes have flown over the region, geophysicists want to know more about the atmosphere, radio conditions, and the earth's magnetism at the top of the world."[51] Edward L. Phillips, Newnan, Georgia, was flight engineer on polar flights that carried his WB-29 directly over the ice islands. As the weather planes passed overhead, radio operators were able to contact the American and Russian scientists on the islands. The dropsonde operator on Phillips' flight was born in Russia and was able to talk with the Russians. The scientists told about equipment falling through the ice into the ocean when the islands started to break up.

Long flights over the polar ice tested the stubborn courage of pilots who could not keep the specter of going down on the ice out of their minds. On one flight the No. 3 engine was running hot and backfiring. As the engineer discussed the problem, the pilot was firm in his intent to continue to the mission-credit point before turning back. He told the crew, "If we go down in the ice, you have two choices—go down or bail out." The weather officer was reporting a surface temperature of 65 degrees below zero, with 45-knot winds, so the crew decided to stay with the aircraft if trouble came. The pilot had decided that he would dive the plane straight into the ice pack. "I don't plan on being polar bear food while I am alive," he said. From that point, Phillips lost interest in polar flying.[52]

The Navy was also interested in the polar ice and, on 8 April 1954, a "Privateer," a WW-II B-24 "Liberator," flew from Argentia, Newfoundland, on what is believed to be the first Navy ice observation flight. AG1 John Rodriguez was the ice observer. Two other flights were scheduled for 16 April.[53] Beginning in 1961, the Navy began "down under" ice reconnaissance flights over Anarctica's Ross Sea. These flights were made by the Aerial Ice Reconnaissance Unit of the Naval Oceanographic Office, and regular forecasting of ice conditions began in 1964. The Deepfreeze Weather Center, at McMurdo Sound in the Antarctic, directed flights from Christ's Church, New Zealand and McMurdo. The recon airmen, calling themselves "Polar Prowlers," observed ice conditions for the shipping lanes, for the general meteorological services of the Navy and for all armed services. In later years ice observations decreased as more expertise was gained using satellite imagery. [54]

Hot clouds created by atomic tests were floating through the atmosphere, and this continued to be an international concern. To complete the net for early detection of the radioactive clouds, a flight of the 513th WRS, at Tinker AFB, Oklahoma, was sent to Dhahran Field, Saudi Arabia, for sampling missions to monitor atomic fallout from Russian tests. This project began with an unexpected urgency. In early March 1950, Col. Arthur McCartan, on a flight from Hansford, Washington, was ordered to stop at Ogden, Utah, for a classified message. He was directed to go to Air Weather Service and the Air Force Office of Atomic Testing (AFOAT) headquarters for an urgent briefing. From there McCartan and his crew were ordered to Saudi Arabia to brief the Air Force commander there on plans for an immediate deployment of four WB-29s to fly sampling missions from Dhaharan over Afghanistan. The Air Force commander, in turn, briefed King Ibn Saud since four B-29s flying from Saudi Arabia might cause problems in a volatile political climate.

Two days later the WB-29s, support equipment and personnel arrived. McCartan writes, "I sent a message to AFOAT indicating their arrival and that we would fly the first mission the following day. By return wire, we were directed to fly the first mission immediately—crew rest was waived. So we went back in the air. There was a closed circulation near a probable Russian atomic explosion site, and the concern of AFOAT was that the atomic particles might escape into the general westerly flow before detection. Time was important—an atomic test was imminent—or so intelligence thought."[55] These special flights, code-named "Hawk," were flown from March through December 1950, before McCartan's weather crews were sent to the Pacific to support atomic tests.[56] There was no way to stop the hot clouds, and the sampling missions could only chart their drift in relation to weather movements and determine how quickly the radioactive fallout would settle to the ground or ocean surface. This was a type of radioactive research.

Every new assignment opened up a new world of flight. Flying, maintenance, and living conditions in the desert of Saudi Arabia presented completely new situations to weather pilots. These flying problems were as dramatic as the ice of the North Pole and as dangerous as the typhoons and hurricanes of the Pacific and Atlantic. McCartan describes unusual conditions during the two months he flew from Dhahran: "It was so hot from April through June that maintenance men could not work during the day. Maintenance could be done at night only, and all base support had to adjust to our weird operation. The dust level over Saudi Arabia was up to fifteen thousand feet, and visibility was poor at low sun angles. Inside the aircraft, dust and fine sand would pile up inside the pressurized vents of the WB-29s. A clean aircraft would become a mess in a short time. I couldn't believe those little desert-like sand drifts in the cockpit. Extreme low humidity caused respiratory problems.

"Dhahran airfield was small, and the base commander squeezed his men to make room for our seventy-five weather recon people—sleeping three decks high in the air-conditioned Quonsets. We were grateful for the wonderful support—non air-conditioned quarters would have been impossible. For recreation, men were allowed to bring motor scooters to the Saudi base in the WB-29 bomb bays—that was a bad decision. As sand drifted over the roadways, scooters slid and drivers were injured. Later, when we went to Kwajalein, Naval Admiral King said, 'No scooters!' That was a wise decision." Capt. Robert Moeller (retired as a brigadier general) did not like the admiral's idea of no motor scooters. In a bull session, one night, he proposed that, to get even: ". . . the next supply flight back to Hickam, we'll get a foot locker and fill it with horse manure. Late at night, we'll spread it on the admiral's lawn. The next morning, he will look at his lawn and explode, 'Those damned Air Force guys have now brought horses to the island!'"[57]

John Hug, 53rd detachment commander, in late 1950, took one hundred men and a flight of six WB-29s to Tripoli, Libya, for air sampling off the coast of Africa. His detachment had one aircraft in the air every day for a twelve-hour mission. They played a waiting game until a hot cloud was discovered. The crew then might stay with it for twenty hours until it drifted off the west coast of Africa beyond aircraft range. On long, boring weather missions, especially over open water or desert, weather crews passed the time by looking for landmarks or historic sites. Across the Sahara, one of Hug's crewmen remembered the crash of "*Lady Be Good*," a World War II B-24, which overflew its Tripoli base. After the crew had bailed out, the plane crashed in the desert.

Years later, the wreck was discovered, and all but two bodies of the crew were found. Although the plane was broken in half, it was in fairly good condition "as far as a crash goes,"—the radios still worked. On a mission flying near the crash site, Hug dipped low and his crew located *Lady Be Good.* A wrecked aircraft, viewed from the air, is a sobering sight. After bail-out, the B-24 crew had tried to walk out, and a diary reported that the men had gone seven days without food and four or five days without water. They probably would have reached safety if the daytime temperature had not been over 100°. Their bodies were well preserved in the dryness, and, years later, they were easily identified. In Tripoli, at air base headquarters, a propeller from the crashed plane was made into a memorial for the lost crew.[58]

With natural weather problems and atomic debris floating around the world, weather crews were kept as busy as at any time during their wartime reconnaissance. Every mission seemed to be a new experience. From the extremes of Arctic cold, snow and ice, to the opposite extremes of desert heat, sand, and visual phenomena, aircraft and weather crews were pushed to their limits. Space exploration would add another dimension.

5

More War, Hot Clouds,
Black Fog, More Weather

In 1950, four and one-half years after Nagasaki, war came back. The United States had removed World War II occupation troops from Korea in 1949, after free elections were held in South Korea. However, fighting continued and forced the United Nations (UN) to ask for troops to restore peace in early June 1950. Then, on 25 June, the bloodiest war to that time, began when communist North Korea invaded South Korea. Immediately President Truman ordered the Air Force and Navy to support the South Korean government with ground and air strikes against the northern armies. Within twenty-four hours of the beginning of hostilities, the 512th (which became the 56th WRS in January 1951) began daily synoptic missions called *Buzzard King* and *Buzzard Kilo* that continued throughout the war. The 512th was stationed at Yokota AFB, north of Tokyo, and the recon flights, beginning in northern Japan, flew three thousand-mile tracks over the Sea of Japan, the East China Sea, and the mountainous Korean peninsula to within one hundred miles of Russian Vladivostok. By July, the 512th was flying two missions a day in WB29s. Their reports were so useful that no ground weather forecaster or commander who sent bombers over Korea could operate efficiently without weather reconnaissance.

Korea was yet another "weather world." Bombers found some of the meanest weather of their experience. This new weather was not viscous like typhoons or ice storms, but soft, benign low cloud ceilings and low visibility which made bombing almost impossible. Korea was on the edge of the extreme weather of China and the jet streams over Japan. Winds aloft were so hard to predict that bombing rendez-vous were off an average of sixteen minutes early or late.[1] Early in the war, weather planes took anti-aircraft fire on about every third mission, unidentified aircraft followed them, and the Koreans jammed their dropsonde signals.

At the beginning of the war, weather aircraft were unarmed. In fact, the only armament section of the 512th consisted of one officer and three enlisted men who administered the issue and use of personal handguns for crew members. Weather aircraft needed guns. An emergency requisition was put through and this request brought one pair of twin 50-cal. machine guns that could be mounted in the tail turrets of the WB-29s. These guns were changed from plane to plane as missions were flown over Korea, and one of the squadron's three gunners flying the "stinger position" was the only protection from enemy fighters. Later, more guns arrived, but with missing parts. So, even after each aircraft had a set of guns, the missing parts had to be switched from plane to plane as missions were flown.[2] This situation spoke for the low "priority" of weather crews in war.

Sending poorly armed weather aircraft into the Korean war zone was a major concern of commanders although a "quirk of war" soon showed that weather flights were not as vulnerable as first thought, a fact that had been learned in the North Atlantic recon flights. In Korea, weather observations were sent "in the clear" to save time and to avoid errors in encoding and decoding. Sending in the clear proved to be good protection since weather information being transmitted was as useful to the Koreans and Chinese as to the U.N. forces, and few weather planes were challenged.[3]

WB-29s were not the only weather aircraft in Korea. A small 6166th Tactical Weather Reconnaissance Flight, flying RB-26Cs, began "snooper" missions in February 1951. Three missions each day were flown over the Yellow Sea and North Korea. However, because the RB-26s had little self protection, were slow, and had a limited ceiling, their missions declined to route patrols providing weather reports along bombing lanes and air support for fighters.[4]

The North Pole missions continued. On 12 August 1952, Capt. Pat Bass, flying a 357th WRS WB-29 No. 44-62197, made the five hundredth *Ptarmigan* mission.[5] Listed as crew for this historic flight, in addition to pilot Bass, from Chandler, Arizona, were Capt. Eugene C. Murphy, Escondido, California; Capt. Sterline R. Funk, Cleveland, Ohio; Capt. Bennett O. Moyle, Big Bend, Wisconsin; Capt. Frank P. Weil, Passaic, New Jersey; Lt. Benford B. Harris, Wesson, Arkansas; Capt. Ben H. Nicholls, Beckley, West Virginia; Sgt. Leroy L. Jackson, Trenton, Missouri; Sgt. Fred W. Joiner, Miami, Florida; Cpl. Billy Cook, Walters, Oklahoma; Sgt. Elba W. Miller, Jr., Charleston, West Virginia; and Sgt. Ennis A. Hilbrant, Phoenix, Ariz.[6]

All is not always serious on "normal" missions. Crews can be cruelly "playful" when threatened by non-flying brass, such as representatives from federal agencies, Congress, the Pentagon, or Air Force headquarters, who often flew on missions to get first-hand information. Navigator Eugene Murphy tells of a potentially serious but humorous situation on a 58th WRS *Ptarmigan* mission in 1952. On this flight was a retired, non-rated colonel, serving as a member of an Air Force committee studying a Congressional cost-cutting proposal to eliminate or reduce "hazardous duty pay" or aircrews' flight pay. Unfortunately, such high level representatives never have a chance to make objective observations—a story that happened over and over in the military. Straight-faced crew members swear that these events are historical fact, happening exactly as related without preplanning.

On this observation flight, the colonel carried his own pearl-handled revolver. Somehow, a fire developed in the cockpit, and the pilot declared an emergency and called for preparations to abandon the aircraft. In the movement around the aircraft, and "chuting up," the colonel's revolver was lost, and he refused to leave the aircraft until it was found. But the emergency was not long lasting. The fire was attributed to a coffee pot and quickly put out, and the missing revolver was found behind the "P" can. The pilot resumed course and continued the unusually long 18-hour flight. The colonel, perhaps not as naïve and unsympathetic as thought, seemed impressed and said, "as far as he was concerned, hazardous-duty pay was well deserved."[7]

With the confidence that comes with skill and experience of weather crews, flying tropical cyclone missions was considered to be relatively safe. However, on 26 October 1952, a weather recon aircraft, making a low-level penetration of Typhoon *Wilma*, went down in the sea three hundred miles east of Leyte. Ten crew members of the 54th WB-29, were lost.[8] Even in peace, weather reconnaissance crews who penetrated tropical cyclones that other pilots avoided, deserved the hazardous-duty pay they received.[9] To prove this point, Leon Attarian, 54th WRS weather technician from Glenmont, New York, describes a scene in an operations office in Guam.

A typhoon was threatening Guam and all aircraft were grounded. The terminal was mobbed with flight crews and passengers trying to get off the island before the storm hit. The operations officer was surrounded by pilots hoping to get clearances, but no flights were being allowed to take off. Into this chaos walked Capt. Kent Thomas, who asked for a flight clearance, and the ops officer quickly signed it. A colonel standing by wanted to know why "this captain'" was granted a clearance when he wasn't. In a loud voice, the operations officer announced, "This is a crew of typhoon

Cpl. Kenneth D. Johnson, 54th Weather Recon Squadron, from Nashville, Tennessee, was honored for his insignia design for his squadron. Cpl. Johnson received a silver cigarette case and an engraved emblem. *USAF photo by T/Sgt J. Sisley, courtesy of Paul S. Betchel*

chasers who are going out into the storm to collect weather data. If anyone of you will take their mission, I will be happy to approve your clearance." All heads turned toward Captain Thomas with "no thanks" comments, and the crowd parted to allow him to pass.[10]

There has been no air war fought and certainly no war won without
crew chiefs and flight engineers who could, almost by magic, find non-
existent parts needed to keep their planes in the air. The unwritten story
of any military service is the extreme daring, deception, guts, and sheer
heroism displayed by the enlisted guardians of their aircraft, as they worked to
keep equipment in operation. Many missions could not have been flown
had it not been for the resourcefulness—skills at "moonlight requisition-
ing"—of the men on the line, most often because of friendship and trade
offs with supply sergeants (and officers) everywhere. We never questioned
where those generators, those starter motors, that new set of spark plugs,
or, even, where a complete engine came from. These parts that had long
been on back order from squadron or depot maintenance suddenly "appeared." Example: high over the CBI Hump on one recon flight, we were in an
ice storm trying to hold altitude. I smelled something delicious. Flight Engineer John Terrell was making soup in a new hot cup that he had plugged
into the electrical system somewhere. "Don't ask," he said. The enlisted
maintenance men were the magicians of the air and on the flight line, and
this is my tribute to them. Don't let the generals know (secretly they did
know), but the enlisted maintenance men won the war by bending and
breaking regulations as need demanded. Try as I might, I can't find anything wrong with that.

One particular story of "I wonder how it happened" illustrates a long-
standing military practice and has been a life-long obsession with Robert
Mann, Shelby Township, Michigan. Mann is president of the Pacific Air
Weather Squadrons Association, who call themselves the *"Pacific Trackers."* Here is his story that has disturbed him for the past forty-five years:
The Korean War started on 25 June 1950, and the 19th Bomb Group departed Anderson AFB, Guam, for Okinawa in the pre-dawn hours of 27
June to fly combat missions out of Kardena. After the 0800 roll call on the
27th, Mann and his crew of the 514th/54th *"Fireballs"* rode to the handstand and received the surprise of their lives. They saw their plane, WB-29
44-86267, with a jack under the tail and gaping holes where the four engines had been! The 19th had flown off to war in Korea earlier in the
night, and Mann thinks—in fact, Mann swears—that after taking all of
the spare parts and engines from the base shops, they decided that they
needed additional engines. So, before daybreak, a team descended on 267,
removed all four engines, loaded them into bomb bays of their airplanes

"The Royal Order of Typhoon Goons" was awarded to all members of typhoon crews who flew into the tropical storms. Here is shown a 54th Weather Recon Squadron certificate to Capt. Carel T. Humme, presented in 1960. *Photograph courtesy of Carel T. Humme*

and were on their way to Okinawa by sunup." And the 514th WB-29 aircraft sat for three months waiting for new engines.

Inquiries to 19th Bomb Group personnel as to what happened to the engines brought blank looks or, "I heard about it but I was not there." Years later, as Mann continued his obsession to find out about the missing engines, he got a (sort of) positive response from a Maj. Donald Jelkes, of the 19th, that Jelkes "thought we told them to take them." Later as Mann was reviewing "official" histories of the 514th, he found the following data for aircraft 44-86267:

Month	Hrs IC*	Hrs NIC**	Hrs AOCP***	Hrs Flown
July	96	648	0	31:55
August	192	552	72	79:35
September	24	696	456	15:50

*In Commission
**Not in Commission
***Out of Commission for Parts

What Mann saw was an official report that his airplane had been in commission 312 hours, had flown 127:20 hours and was out of commission for parts only 528 hours. "I know the *Fireballs* were good, but how could an airplane without engines be in commission 312 hours and fly 127:20 hours?" There is a time-honored answer to this mystery, as old as history itself. Much of military history is heavily weighted by ego and the human desire to "make my outfit look good." This military phenomenon is expressed with a special term: "CYA,"[11] and squadron histories often reflect a strong emphasis on CYA.[12]

Attempting to look good, as weather units competed for attention, jealousies increased, squadron zeal surfaced, and commanders fought bravely to pull their units out of the shadows of favored and more "politically correct" units. In the 1950s, the darling of the Air Force was the Strategic Air Command (SAC). At a recent Air Weather Service (AWS) meeting, at Tucson, Arizona, retired Gen. Robert Moeller confessed his 57th Weather Squadron's resentment of SAC: "Every time SAC flew a B-29 one hundred hours, the *Air Force Times* would run a 'stop the presses' big story and a front-page picture. So, in 1953, at Hickam, we got enough of that. During one month we flew the same WB-29 366 hours and 20 minutes in one month. Proudly, we painted a big "366 hours" on the aircraft's nose, took a picture and sent it to the *Air Force Times.* Nothing! The *Times* and no other paper ran the story or the picture." General Moeller has carried this burning coal into retirement.[13]

The Strategic Air Command (SAC) was a runaway power in the Air Force, and its enthusiasm often exceeded reason. During war SAC was dedicated, and in training games after the war, it was just as serious. Marshall Balfe, Santa Rosa, California, copilot of the 57th/513th WRS, was flying from Hickam, Hawaii, to California, with a full B-29 crew. Their destination airport was closed with fog, and they were cleared by traffic control to their alternate airport. They landed, taxied in, and shut down the aircraft on the ramp. Suddenly, a jeep full of soldiers with guns challenged the crew, demanding to know who they were. Balfe wrote, "I spoke up first. See what is written on the side of our airplane—'U.S. Air Force.' That's who we are." Nonetheless, the crew was herded into a small room in the

When 57th Weather Squadron Commander Robert Moeller set a record for a WB-29 flying 366 hours and 20 minutes in one month, he had the number painted on the side. His enthusiasm and the squadron record went unnoticed. *Photograph courtesy of Gen. Robert Moeller, USAF (Ret)*

operations building. After some time they were returned to the airplane and told to leave. "We had landed at a SAC base when the troops were playing games—we happened to innocently land at the wrong base at the wrong time and were considered enemies."[14]

There were many dedicated people, from crew chiefs to commanding generals, worthy of a special mention in weather reconnaissance. However, one man earned a special place in recon history. A tribute is due to Brig. Gen. Richard "Dick" Ellsworth, a pioneer in weather recon and one of the most highly trained and experienced meteorologists in the Air

Gen. Richard Ellsworth, in attempting to improve weather reconnaissance and test radar defenses, lost his life when his B-36 crashed into a mountain in Newfoundland. The B-36 was the Air Force's largest aircraft and was a major deterrent in the cold war. Ellsworth AFB, Rapid City, South Dakota, is named in his honor. *Photograph courtesy of Bernard A. Shea*

Force. Ellsworth's life was dedicated to the idea that weather reconnaissance was important to defense, and his death came as he was trying to improve the defenses of this country. General Ellsworth was working within the Strategic Air Command (SAC) and its RB-36 program to build and maintain a weather recon superiority.

Ellsworth's weather career began in 1941, as an Air Corps meteorologist. In July 1943, he was assigned to the China-Burma-India theater as regional control officer of the Tenth Weather Region and commanding officer of the Tenth Weather Squadron. In the CBI he flew four hundred combat missions in China and over the Hump, flying his beloved C-47 "Betsy." He helped pioneer night flights over the Hump and establish a weather forecasting and recon network in communist-held China to support B-29 air raids against Japan.

After the war, in 1946, Dick Ellsworth left headquarters of the Air Weather Service to become commander of the 308th Weather Reconnaissance Group and pulled together various air squadrons to form the focal point in the organization of weather reconnaissance. He established weather recon in the Pacific from Fairfield-Suisan AFB, California, pioneering such projects as the *Ptarmigan* recon flights over the North Pole. After graduation from the Air War College, he was assigned to the SAC at Barksdale AFB, Louisiana, and, in November 1950, was sent to Rapid City AFB, South Dakota, to work with Gen. Curtis LeMay, directing weather operations and flying the huge six-engine B-36 bomber.

Ellsworth's job in developing a weather recon capability in SAC was not an easy one. Commanding Gen. Curtis LeMay, with his RB-36 program, was not dedicated to weather reconnaissance as a single-purpose mission. Colonel (at that time) Ellsworth was commander of the 28th Wing with Lt. Col. Charles Markham, staff aerial weather reconnaissance officer. Together they convinced General LeMay to integrate a weather reconnaissance capability into its strategic functions of photo, electronic countermeasures (ECM) and radar reconnaissance. Markham with many years of reconnaissance experience, tells the story of Dick Ellsworth:

"When I first got to the 28th, no one was happy to see me—only two people had the slightest interest in weather recon: Wing Commander Colonel Ellsworth and me. We had the whole show to run. Worst of all, Ellsworth had SAC commander General LeMay to keep happy. Four-star General LeMay limited himself to ten minutes of happiness a day—he was a hard man to please. It was said that he kept colonels pilloried in two parallel rows outside his office and blew cigar smoke in their faces as he walked by. We had a big sales job to do."

With the anti-weather feeling in SAC and the difficulty of working with General LeMay, the best contribution Ellsworth could make was to convince the general not to work against the recon effort. Midway through Markham's tour at Rapid City, on 5 September, Colonel Ellsworth, at the age of 41, was promoted to brigadier general. On his return flight from the promotion ceremonies, the new general stepped off the plane with horns blaring, drums pounding and the band marching ahead of his car. The whole base was proud and happy. Best of all, weather had its own general.[15] The new general and his weather staff, with persuasion, logic, and quiet diplomacy, began to develop the weather recon program in the 28th bit by bit. He added weather training to various crew position requirements in the RB-36. So, we had nose-gunner-weather observer, navigator-weather observer, and a dropsonde capability was added to the RB-36—the dropsonde operator had to be trained as a gunner because that was the position replaced. "Before long, everybody accepted weather reconnaissance as a part of the wing's new job."[16]

General Ellsworth went to his death trying to improve the radar defense of the nation which was controlled by the Air Defense Command (ADC). In an intense rivalry with ADC for defense primacy, SAC set out to prove that ADC's defense net could be penetrated. On 18 March 1953, Ellsworth planned a low level penetration of ADC radar along the North American continent. Secretly, hidden by weather, Ellsworth's plan was to fly through a front under ADC's radar and appear in the clear at thirty thousand feet above Dayton, Ohio, for a simulated bomb run. The mission required radio silence and no aircraft radar use. He left the Azores in an RB-36 and navigated toward the nation's east coast, flying low under the radar net to test the coastal defenses. A freak weather pattern in the North Atlantic caused a serious error in calculation of ground speed, and Ellsworth and his crew, thinking they were over water, crashed into the hills of Newfoundland in the fog at night. Twenty people were killed instantly. Thirteen more were killed when a rescue plane crashed on takeoff.

On June 1953, Ellsworth Air Force Base, Rapid City, South Dakota, was dedicated by President Eisenhower in honor of Brig. Gen. Richard Ellsworth,[17] the only major air force installation named after a weatherman.

The Convair RB-36 was a pusher aircraft with six propeller engines and four jet engines, dedicated to weather recon within the SAC fleet. The B-36 was the nation's atomic strike force during the cold war of the fifties. Its only advantage was its 9,500-mile range. With its 230-foot wing, 163-foot length and 46-foot height, the B-36 flew at 340 MPH at thirty-five thousand feet. It probably satisfied those who thought bigger was better, but it never had a chance to prove itself in wartime, and, perhaps, its greatest contribution was in weather reconnaissance. Weather observers first rode in the nose position in the RB-36 with duties that required observing, recording, and transmitting weather observations in code to using agencies. Later, the observers were moved to the rear of the aircraft to take and transmit dropsonde observations. Observers doubled as gunners and scanners. From January 1950 to June 1958, Bernard A. Shea, Rancho Cordova, California, who provided information for the RB-36 weather reconnaissance, flew 3,492 hours on weather missions in the RB-36. Shea was interested that the SAC RB-36s become a part of air recon history.

In 1954, Gen. Thomas Moorman, commander of the Air Weather Service, was concerned that the weather service was not getting enough good applicants in meteorology, and he discussed this in a staff meeting. Maj. William C. Anderson, an energetic public relations officer and weather recon pilot, decided the best plan to get publicity was to "start at the top."[18] He called Edward R. Murrow at the Columbia Broadcasting System (CBS) and asked if he would like to fly through a hurricane—Hurricane *Edna* was prowling the Atlantic and was the top news story at that time. Before he could get an answer, Murrow's boss and producer, Fred Friendly, called and in most unfriendly terms, explained to Major Anderson that hundreds of people depended on Murrow's "See it Now" show and if "anything happened" to Murrow in the hurricane they would all be out of jobs. "I don't want him to fly through any hurricanes," Friendly said, and instructed Anderson to tear up Murrow's phone number.

However, Murrow did accept the invitation, and he and his camera crew flew to Bermuda and, on 10 September 1954, in WB-29 4469987 of the 53rd WRS, Murrow flew through Hurricane *Edna* as it nipped Cape Hatteras, North Carolina and paralleled the eastern coast to above Maine. The aircraft commander was Capt. Francis E. Wilson, and the pilot was Capt. Charles E. Fleischer. From this flight and the "See it Now" television show, came Murrow's famous ". . . flying is made up of many hours of boredom, interspersed with a few minutes of stark terror,"[19] and the most often quoted "The eye of a hurricane is an excellent place to reflect on the puniness of man and his works. If an adequate definition of humility is ever written, it's likely to be done in the eye of a hurricane."[20]

On the ground, after the flight, Murrow slapped Anderson on the back enthusiastically, "What an exhilarating experience!" After the "See it Now" telecast, the weather service was "literally engulfed" with applicants for training as meteorologists and weather reconnaissance crew positions. In time Brigadier General Moorman was promoted to two stars, and weather recon received its greatest boost.[21] However, it is fair to say that General Moorman received his promotion and eventual appointment to the superintendency of the Air Force Academy on his effectiveness as a commander and not necessarily as a result of the Murrow-*Edna* flight. The only "casualty" of the *Edna* flight was Dr. Robert Simpson, director of the National Hurricane Center, who had planned to launch some balloons from the weather plane. In his words, the experiment was preempted "by that well-known newsman Edward R. Murrow and his film crew. I was a bit teed off."[22]

Air crews were accustomed to unusual problems in the air. However, as they flew into previously inaccessible areas of the world, crews discovered unknown situations about operations, navigation, and crew perceptions in extreme climates. Pilots encountered visibility problems unique to the Arctic, and later, on flights into the desert, they found many of the same problems. Strange visual illusions preyed on the pilot's perceptions—refraction phenomena, mirages, ground haze, and distortions caused when light seemed to pour in from every direction—from snow or over water—a type of visual vertigo. Pilots called this world of no-shadow whiteness, when the sky and ground snow or haze blended together, "flying in milk"—a surreal intense brightness not found in any other instrument flying. Crews were also plagued by "distance distortion." When they had good visibility, objects at great distances seemed larger—like the moon coming up in the eastern sky! Navigators were often as much as three hundred miles in error estimating the distance of land masses. Primitive lore from ancient sailors, as they navigated over miles of trackless ocean, often explained new experiences. The ancient Polynesian "sky map" lore was rediscovered in the Arctic and over vast distances of open water. A small cloud in a clear sky would show the location of an island, and the lagoon of an atoll would be reflected on the underside of a cloud.[23] In the frozen wastes of the north, a uniform overcast often reflects what is on the ground below long before it can be seen, as every pilot flying over oceans has learned, and rediscovered a lost instinctive link to nature that had been numbed by technology.

Still other assignments were being added to the reconnaissance sampling and synoptic missions. Before the 57th left Tinker AFB, Oklahoma, to participate in the 1951 atomic test *Greenhouse* in the Marshall Islands, Col. Arthur McCartan, commander of Flight AF 3.1.3, two of his aircraft were equipped with airborne magnetometers to map the earth's geology from the air on a special project that lasted three years, including a program to chart Canada. The aircraft also carried pyroheliometers to measure the heat loss or gain from the earth's atmosphere at the planes' flight level.[24] These were special one-time projects, and were a reflection of AWS commander Gen. Donald Yates' interest in scientific projects, and his belief that as long as aircraft were flying they could carry out additional weather or geophysical projects. The 57th's main mission, however, along with other weather squadrons in the Pacific, was monitoring the atmosphere over a million square mile area in the Pacific, not only for atomic bomb testing, but for any thermonuclear activity underway in Siberia.[25] As long as atomic tests were being conducted by prickly nations, weather's radioactive sampling missions were becoming more urgent. Atomic tracking data at the Pentagon showed that atomic debris would circle the world three times and weather officials were convinced that it was the tracking missions that kept weather recon alive.[26] Air samples would indicate radioactivity in the atmosphere, and laboratory testing could determine types of bombs and even where they might have been detonated.

In flying through the hot atomic clouds, weather crews faced an invisible enemy. WB-29s became contaminated and had to be cleaned before they could be used again, and before landing, pilots tried to fly through rain showers to help clean the aircraft. On the ground, planes went through wash racks, and crews had showers and new clothes. Usually, no contam-

Flying radioactive atmospheric sampling missions was a lonely and dangerous assignment for air weather crews as they sought out and tracked an invisible enemy. Here, a 59th Weather Squadron WB-29 follows a "hot cloud." *Photograph courtesy of Lester Ferriss*

ination inside the aircraft was found since the B-29 pressurizing system had large chemical filters that stopped debris from getting inside although a crew member might "get dirty" by rubbing against the outside of the contaminated aircraft.[27] Col. Lester Ferriss, 59th WRS aircraft commander, from Salinas, California, remembers, "The decontamination of aircraft went on all night. . . . I can still visualize the garish flood lights, fire trucks and streams of water . . . scores of men with mops and brooms, climbing over and scrubbing the big silver planes. . . . It was memorable and [during atomic tests] I don't recall how long it went on—ten days, two weeks, three weeks—but I do know it strained our resources, both equipment and people."[28]

Based on research to date, the average radiation dose received by approximately two hundred thousand Department of Defense test participants was about 0.6 rem. The current federal guidelines for radiation workers permit external exposures of five rems per year.[29] Today, in 1996, there is concern that air crew members involved in radioactive testing and atomic aerial surveillance, in the 1948-1958 missions, may have received too much radiation. The incidence of cancer and other radiation-associated diseases is now becoming a source of concern. Robert A. Mann, president of the Pacific Air Weather Squadrons Association (PAWS), calling themselves *"Pacific Trackers,"* is conducting a campaign with nuclear agencies and the Department of Veterans Affairs to recognize the problem and conduct testing and treatment of air crew members possibly affected.[30]

Radiation did not affect the basic operational problems of aircraft, but B-29 weather crews constantly fought routine mechanical problems. The main and most often mentioned problem reported by flight crews was the fuel transfer system which delivered fuel from a series of tanks to the four engines. The B-29 system allowed any tank to supply any engine or all tanks to be fed into a central manifold that serviced all engines—a plumber's puzzle. First Lt. Donald Shertz, weather recon officer of the 57th, reports a sphincter-straining situation on one flight, "We were climbing past seventeen thousand feet when all four engines suddenly quit. The engineer, bless his soul, immediately suspected the automatic mixture control had thrown a full-lean signal to all engines, cutting off the fuel supply. The pilot immediately sent a MAYDAY signal and, with props windmilling, we were gliding quietly toward the sea. This alert engineer immediately began dismantling the automatic mixture control and discovered a loose washer had shorted the output signals. He removed the washer and turned the power back on, the engines caught, and the huge plane leveled off at twelve hundred feet. The pilot canceled the MAYDAY and we completed the mission—I wish I could remember the name of that superb engineer."[31]

To hear engines sputter and quit when all fuel has been used is a pilot's private dread, but to have the fuel gauges bumping "E" when you know there is a tank full of fuel that can't be used is another kind of agony. Lt. Chester Stone, pilot for the 514th WRS, from Cypress, Texas, and his WB-29 crew were flying from Anderson, Guam, to a point six hundred miles north of Tokyo, with the mission ending at Yokota AFB, Japan. The ceiling was two hundred feet, visibility two miles, with freezing temperatures near ground level. As the mission was nearing completion, the Yokota range suddenly went off the air leaving the crew holding a bag of very bad weather with no guidance. The radio compass had also failed. Pilot Stone decided to fly DR[32] to Yokota and, after a radar approach, trust GCA to take them in. This plan worked, and a safe landing was made in almost zero visibility. Next day, on the return flight to Guam, after passing the point of no return, pilot Stone found that the fuel transfer system had failed and there were 550 gallons of desperately needed gasoline trapped in the bomb bay tank that they could not get to. The remaining fuel in the main tanks would have to be stretched to the limit. As they desperately tried to raise an alternate airfield on the radio, they found all operations had shut down at midnight. There was nothing left to do but try to reach Guam at long-range cruise.[33] Approaching Anderson, there was too little fuel to make a standard instrument approach—all gauges were registering empty—so Lieutenant Stone maneuvered the huge plane on instruments for a straight-in approach to the field. At six hundred feet they broke out of the overcast and landed on Anderson's eight thousand-foot runway. The engines coughed and died as the plane was being taxied from the ramp to the handstand.[34]

Bermuda was a plush assignment for 53rd WRS personnel and their families, and although it might be unfair to say that their main worry was the number of golf balls driven into the ocean, Bermuda was a "country club" base, and everyone considered themselves lucky to have the assignment. However, comfort stopped at the flight line. Recon missions were flown into some of the worst storms experienced by any squadron in the weather service. Fear and possible death rode on every flight and was a constant drain on squadron morale. Routine flights even in good weather often brought terror to the most hardy. On one test hop out of Bermuda, a WB-29 had a runaway prop. A prop out of control is an extremely dangerous situation, since the huge 16-foot propeller, spinning wildly, does not respond to any automatic or manual control, usually can't be feathered, and the RPM goes so high it sounds like a freight train as the tips pass the speed of sound. Vibration is unbelievable. Even when the engine is shut down, the huge propeller, still spinning, creates so much drag it is often impossible to control the aircraft. On this one mission, miles out over the ocean, with both pilots struggling, the crew was preparing for bailout. They had just gotten the nose wheel door open when the aircraft commander decided that by reducing speed to 130 MPH, they might get back to the base. Not much was said as the plane limped home, but thoughts were heavy. The navigator on board had five or six children. He was so frightened when they landed, he didn't say a word to the pilot, walked straight to operations about 150 yards away, took his 'chute off, threw it on the floor, said, "I quit!" and walked out. He did quit, taking his wife and family back to the States on a ship.[35]

Eleven weather crewmen were lost on the evening of 3 November 1949, when a 373rd WB-29 crashed during an emergency landing attempt at Kindley Field, Bermuda. Two crew members died later, and two had only minor injuries. The WB-29 had taken off from Kindley enroute to Puerto Rico to investigate a closed circulation in the South Atlantic, near British Honduras. Shortly after takeoff, the plane developed an engine oil leak, and after feathering the No. 1 engine, the crew asked for permission to return. On the approach for landing, the plane, without explanation, turned to the left, crossed St. George's over the swing bridge, just missed a hill above Mullet Bay, and crashed into the sea about three hundred yards from shore. This was the worst air crash in the history of Kindley.[36]

Another incident, still a mystery to many Bermuda crews, was a strange accident on a weather mission. The assistant operations officer received a report that a man had fallen out of the front bomb bay of a WB-29 soon after takeoff. There was all sorts of speculation about what happened. John Hug reports, "We received word about fifteen or twenty minutes into the flight, telling us that this man had fallen out, jumped, or whatever. The ocean was very rough, and there was so much wind when Air-Sea Rescue got there, they dropped a raft and it just tumbled and tumbled. All the crewman had was a parachute and a one-man dinghy, and he must have taken in a lot of salt water. He was rescued and stayed in the hospital for a day or two, and walked out smiling."[37] This story is told from another source about Capt. Wallace Taylor, the man who fell out of the WB-29: ". . . the aircraft was at seven thousand feet when a crew member smelled gasoline in the bomb bay. Captain Taylor climbed into the bay to look for the leak. His flying suit accidentally caught in the emergency bomb bay release, the doors flung open, and Taylor went tumbling into space. He opened his parachute and, when he hit the water, inflated his one-man raft. Taylor bobbed around for six hours before rescue boats, battling 30-foot waves, were able to pick him up. He was not injured but was violently sea sick."[38]

Bermuda was the home base for the 53rd WRS, the "*Hurricane Hunters*." Since the squadron's beginning in 1944, everyone was enthusiastic about hurricane reconnaissance. In a burst of pride, Dr. Robert Simpson, director of the National Hurricane Center (NHC) from 1968 to 1973, in a video documentary, wrote, "The 53rd '*Hurricane Hunters*' scientifically discover the birth, measure the growth, probe the strength, observe the movement, and eventually record the death of nature's gangster sons."[39] The NHC tried new approaches to storm recon. During the summers of 1952 through 1954, Dr. Simpson, in hurricane research with the 53rd's Flight A, successfully launched weather balloons from the bomb bay of a WB-29 into the eye of the hurricane. These were constant-level balloons so they would float in the center of the eye and could be tracked—one balloon stayed suspended in the eye for twenty-four hours and followed the storm's movement during that time. This balloon procedure helped the recon crews in planning missions.

On 18th September 1953, the 53rd suffered another tragic accident. As Hurricane *Dolly* was headed toward Bermuda, all aircraft not on missions were evacuated to Florida. One WB-29, No. 62277, lost an engine but was able to make it safely to Hunter AFB, Georgia. A new engine was installed and, after a flight test, the pilot filed a flight plan to Bermuda late in the afternoon of 18 September. About five hundred miles east of Cape Hatteras, at twenty thousand feet, the No. 3 propeller tore loose and struck No. 4 engine. Almost immediately, the entire right wing was in flames. As the plane entered a flat spin, a bail-out order was given—there was no time for a "MAYDAY" message—and nine of the crew were able to bail out. All of this happened at sundown. The plane crashed in flames into the ocean a few hundred miles off the Carolina coast. Seven men were lost, but the nine who bailed out survived. A/1c Norman Prosser, Kansas City, Missouri, said, "I was falling, on fire, and thinking I was dead. I pulled the rip cord and, as the parachute opened, an engine fell past me." In the water, with two other survivors, Prosser fought off sharks during the night, and all were rescued the next day.[40] When the aircraft was overdue at Kindley, a massive air and surface search was organized. Lt. Col. Don Offerman, New Braunfels, Texas, flew fourteen hours on the search mission and found the survivors. A Coast Guard patrol plane landed near one of the survivors but could not take off because of the huge swells. The plane was lost, but the Coast Guard crew and WB-29 survivors were picked up by a cruise ship and treated as heroes on the way to New York City. Colonel Offerman was on the accident investigation board and flew to New York to interview the men.[41]

In November 1953, the 53rd WRS moved to Royal Air Force Station, Burtonwood, England (later, to Alconbury, and eventually to Mildenhall AFB) to begin a seven-year stay in the United Kingdom. Crews from England flew *Falcon* reconnaissance tracks over the North Atlantic and North Sea as far north as the Arctic Circle and south to the Azores area. Flight A of the 53rd remained at Kindley Field, Bermuda to fly *Gull* tracks over the South Atlantic and Caribbean. These flights covered the two major storm threats—Arctic blasts from the north and hurricanes from the south.[42]

As the 53rd was assigned to England, weather crewmen prepared for a royal, steeped in tradition atmosphere, wholesome and very proper. The British, although often difficult to work with, were respected for their skill. A Royal Air Force pilot and aircraft were a team and were superior in the air. It was said that the British acted as if they owned the world, and the Americans acted as if they didn't give a damn who owned the world. So, to Burtonwood, England, came the seasoned weather crews of the 53rd, and it was difficult to imagine flying conditions that would seriously challenge their skills.

Eugene Wernette, aircraft commander, remembers England for FOG—black fog, pea soup fog, cold fog, warm fog, supercooled fog—bloody FOG. "We were based at Burtonwood, halfway between Liverpool and Manchester, so, regardless of weather conditions, we had industrial fog from one or the other. Sometimes it was pitch dark at noon and would stay that way. We trained crews to fly 200-ft ceilings and one-half mile visibility on a regular basis. Eighty percent of the time our takeoffs and landings at Burton-wood were under minimum conditions—all IFR (instrument flight conditions) and we almost always had GCA (ground-controlled approach) guided landings. Any guy who flew out of England could call himself a fully qualified pilot."[43]

Navigation in England was precarious to pilots and navigators trained in the States—"England presented nothing more than a patchwork crazy quilt . . . none of the straight roads and fences, no space between towns, no high tension lines to follow. We had a hopeless conglomeration of winding roads, vari-colored fields, an occasional railroad and always five-mile visibility. When visibility reached ten miles, the group weather officer declared that a meteorological phenomenon existed." [44] Crews suffered weather and navigational shock and "jolly old England" was not so jolly after all. Pilots were more concerned with the problems of taking off and landing at their English bases than the weather encountered on the missions.

In the fall of 1954, another flight of the 53rd WRS was sent to Saudi Arabia, to be stationed in Dhahran, and fly *"Hawk"* tracks over the Gulf of Oman, to Lahore, Pakistan and return, and synoptic flights to the North Pole. Because they flew over the pole, children of the 53rd crew members persuaded their fathers to take letters to Santa Claus and drop them over the pole. Soon the word spread and letters from all over Britain were carried. By the Christmas of 1955, letters from all over Europe poured into the weather squadron.[45] The same was true in the Pacific. The 55th WRS, flying *"Loon Hotel"* missions from Hickam AB, Hawaii, dropped letters in Fairbanks, Alaska, to be postmarked and mailed from North Pole, Alaska. This Christmas "postal" service was publicized in the Fairbanks newspaper,[46] and attracted world-wide attention.

As the cold war wore on year after year, atomic explosions continued to worry Washington, and weather recon aircraft began to receive unusual secret missions, always on extremely short notice. David G. McFarland, Palm Beach Gardens, Florida, was a weather officer in the 54th WRS, based on Guam. In 1955, after completing a season of twelve typhoon eye fixes, his crew was alerted and selected for a special mission. After a flight to Yokota, Japan, the crew, without explanation, was issued heavy—very heavy—flight gear, and a secret briefing was held. In an article in *Weatherwise,* McFarland, described a secret mission:

Behind a locked door and lowered blinds, two colonels, two air police, and three men in fatigues with no insignia, began the briefing. One colonel said, "Gentlemen, you are going to fly a top secret mission to gather samples for the radioactive air mass from the detonation of a hydrogen bomb. You will wear no ID—only dog tags are permitted. If forced down, try to land near one of the uninhabited islands south of Kamachata. If you ditch in the ocean you will freeze to death in two minutes. If you are captured—he paused and tried to make eye contact with all who would let him—and tell the enemy the nature of your mission, you will be court-marshaled if you ever return. The rest of the briefing was blurred, but, in essence, we would fly a normal weather track across Japan, follow the Kurile Islands northeast to a point south of Russian radar and turn west toward Russia. I was to send dummy weather obs as though on a regular weather mission.

"On the third day after the briefing, an assigned crew member from the Atomic Energy Commission (AEC) came in with a box of filters and said that we would 'leave in one hour.' With enough fuel for a fourteen-hour mission, the WB-29 was airborne at 0700. After flying a normal weather track and passing over an F-86 base on the northern tip of Hokkaido, the pilot was directed to thirty thousand feet and a heading of 290 degrees. Brownish clumps below were the islands the crew could abort to. The cold was intense. With heaters at full blast, the cockpit was still cold and the crew's feet and fingers were numb. The outside temperature was -60 degrees C (-76 degrees F) and, below and above, everything was white.

"The B-29 flew in and out of thick cirrus clouds as the AEC 'crew member' called out different headings. After what seemed forever, the pilots were directed to return to their normal weather track. The first navigator said he would do that as soon as he figured out where the hell we were. There was no enemy interference—no MiGs—and we were back on the ground at 2000. The AEC crew member packed his filters, thanked us, climbed into a Jeep and disappeared. No one heard from him or the contents of the filters."[47]

On many of the synoptic and typhoon tracking missions, crews reported Russian MiGs "flying formation" with weather aircraft. Ted C. Jafferis, Overland Park, Kansas, weather recon officer of the 56th WRS, mentioned that one of the hazards of weather flying, in his experience on *Buzzard* tracks, was "having Russian MiGs fly over us and 'lock on,' forcing us to take evasive action like flying close to the ocean or into clouds."[48]

Weather recon crews became accustomed to these super secret missions. They rarely knew exactly the purpose of the mission and were always sworn to secrecy. In recently declassified documents, historians have learned the extent of these missions and the unknown number of U.S. flight crews lost in the icy waters around the Soviet airspace. At a recent reunion of the Air Weather Association, in Tucson, Arizona, I talked with dozens of crew members who told of sampling flights near Soviet or Chinese territory during the cold war. One former radio operator mentioned his contact with an unknown American aircraft high

above the Arctic circle. The pilot refused to give his position or altitude
except to say that he was looking down on the weather plane. Most
weather recon veterans seemed unwilling to discuss their experiences, so
I made no attempt to coax information due to the sensitive nature of their
missions. Sadly, many were seeking information about friends and rela-
tives who had been lost on the missions.[49]

As atomic atmospheric surveillance continued, weather observers and
their crews continued to try to find a way to locate and safely approach
the invisible airborne radioactive debris. The object was to find the edge of
the cloud without entering the hot area by measuring gamma radiation. Radioactive cloud movements depended on invisible upper winds that carried the hot clouds away from ground zero, and the magnitude of the
atomic explosion determined the height of the plume. Conscious of aircraft and crew safety, commanders kept their planes below an altitude of
twenty thousand feet and used a radiation detection instrument (radiac)
to help find the edges of the clouds. New discoveries were often made. For
example, one mission tracked a hot cloud many miles across the equator
and dispelled the theory that there was little or no exchange of atmo-
spheres between the two hemispheres.[50] After some experience with the
movement of radioactive clouds, flight weather forecasts began to include
a guide to where hot clouds were expected to be one to eight hours after
detonation. This guide was called "radax" and was quite successful in
plotting and forecasting movements.

The atomic fires that burn in the basic structure of nature brought
anxiety to the crewmen who searched out and followed the hot clouds.
However, if aircrews can feel fear, they can express pride in their role of
waking a world to the dangers of unlocking the atom. Glenn Bradbury,
weather observer from Fort Worth, Texas, who began his weather career
in 1943 in the North Atlantic, wrote an essay on "Radiological Surveillance," in July 1994, expressing these thoughts:

"The 57th WRS (one of four squadrons flying atomic testing missions)
was honored to have been selected for nuclear weapons testing. To observe a
detonation, to feel the shock wave, to witness the rise of the mighty plume,
and to see the unbelievable destruction of an island presented an experience that transcended awe and caused us to wonder about the future of
Mankind on Planet Earth. I [Bradbury] was reminded of a Biblical reference, "... and they shall beat their swords into plowshares, and their
spears into pruning hooks; ... neither shall they learn war any more; ..."
(Micah 4:3) And, I am grateful to have had the opportunity to participate
in a project of this magnitude."[51]

6

"Pro Bono Publico,"—Recon for the People

During the cold war of the fifties and sixties, the nuclear threat was a gnawing anxiety in people on both sides of the iron curtain. To many the underlying fear seemed to be, "We may all die in an atomic Armageddon." Scientists were predicting a possible nuclear winter with atomic dust and particles blocking out the sun and bringing on a new ice age. Washington was constantly on alert. Responding to this international fatalism, weather reconnaissance squadrons became deeply involved in nuclear research, bomb tests and tracking atomic clouds floating throughout the world.

However, regardless of the importance and glamour of following hot clouds from nuclear explosions, weather crews never forgot that flying typhoons, and knowing their location, was also a priority. Typhoon recon were the real macho missions, and the challenge of the great storms was a source of pride for many crews. James Slaeker, 54th WRS aircraft commander, from Federal Way, Washington, wrote, "Typhoon flights were different, the stakes were higher, and we had the feeling that we were helping people who were in the path of the storm."[1] Another comment from a weather observer also expressed this feeling, "As a career military type, I always had a good feeling about weather reconnaissance because, while it filled a military need, it also contributed to the well being of the general population . . . saving lives and property. Not much in the military budget gives that kind of direct contribution to the well-being of the taxpayer."[2] The idea that flying violent weather as a peace-time public service made dangerous missions easier to fly.

The Pacific is such a vast expanse of often troubled water, it was impossible to know what is happening in every part of the ocean, although weather men were expected to know, and it was hell if someone found a storm the forecaster was not aware of. The life of a forecaster was often clouded by known and unknown storms. In the peace and quiet of the weather station, there was nothing as unsettling as having a colonel come into the map room, screaming, "Where's that typhoon?" It happened in December 1955 in Guam and illustrates the problem in trying to track storms over millions of square miles of open ocean without adequate observations. David McFarland, Palm Beach Gardens, Florida, tells about Typhoon *Patsy* when he was a flying weather officer with the 54th WRS.

"It was early December, and the typhoon season was over . . . after completing eight typhoon eye fixes in our sturdy eleven-year old WB-29. Suddenly, breaking the calm, Lt. Col. George Winslow, red faced and out of breath, burst into the room. 'Where's Boggs?' 'Here, sir,' said Maj. Frank Boggs, weather detachment commander. 'Major, I just got a call from the general [Brig. Gen. Henry Ladd, commanding officer of the First Weather Wing, Japan] who said he had flown through the worst storm he'd ever seen. Know anything about it?' Boggs turned pale. He turned to Maj. James White, typhoon training officer, with a question in his glance. Having faced many typhoons, White was calm in this storm. 'The truth, colonel, is that we don't know much about it. We have a few pilot reports of a possible storm, and forecasters at Clark [Philippines], said it would weaken and turn north. But we'll have a plane in the storm tomorrow—sir!"

The colonel banged the door shut. To lose a storm, or not know its location is a weather man's secret fear—to have to explain to a general that you didn't know is a problem of catastrophic proportions. McFarland describes the next morning's flight, "Preflight was at 0530, takeoff at 0600. Aircraft commander Fred Tucker gave the order to 'load up.' As I climbed the nose wheel ladder, I was always staggered by the first whiff of what I called 'old sweat'—the sweat of every man who had ever climbed that ladder, multiplied by the number of missions the old WB-29 had flown. The smell of missions past stalked me as I walked from the navigator's compartment, between the pilots' positions, to the weather observer's place in the nose of the plane—the best seat in the house! The interior decor of our big bird was camouflage green with chips and scratches."

After a good take-off roll, the huge B-29 bucked at just the right place as it hit the dip in the runway. The nose wheel thumped coming up, followed by the main wheel thumps—all correct and reassuring noises to an experienced crew. The search altitude was ten thousand feet, with the sky and ocean looking dark and angry. The pilot took a heading to the north-

Typhoon *June* moves toward Guam, in 1976, as shown in this radar photograph. Clarence Miller, dropsonde operator, flew several fixes into the storm in the air and then rode it out on the ground as it passed over Guam. *Photograph courtesy of Clarence Miller*

west. On typhoon missions, although the pilot is in command of the plane, the weather officer directs the mission into the eye and out. At 0700, visibility was clear enough to see the water, the flight's position was about six hundred miles from the storm with expected winds of 15-20 knots. Instead, the ocean was in a frenzy—long streaks of foam and gobs of green water—a typical storm. The Navy explains that green water is the turbulent upswelling of sea weed from the depths of the ocean. To be a typhoon, a storm must have a wind speed of 70 knots (75 MPH) or above, and this storm had winds of 100 knots or more! The pilot took a heading to enter the eye from the weaker westerly quadrant.

McFarland was taking observations when a radio message crackled in: "Morning fix aborted . . . severe turbulence . . . approx max winds at wall cloud 175 knots! Enter eye at your discretion." This storm was now Typhoon *Patsy*—she had already earned a name. "Shall we go in?" McFarland asked, looking back at Fred. He said, "Let me check the panels." When the flight engineer reported everything "normal and in the green," the pilot tapped the back of my seat, "Take us in Mac, I won't abort unless we have to." Visibility was zero — the ocean was in a fury no one had seen before; like a white lace table cloth laid over a black table top with huge areas of green ice cream on top. I wrote in the log: "Approaching wall clouds of *Patsy* with estimated winds 180 knots (200 MPH)!" As I wrote, the plane went bump-bump and bang!—we dropped two hundred feet and were in the eye! ". . . wind direction indicated that we had been blown around and had entered the eye at near the most dangerous side. I continued the log, '. . . circling inside the eye waiting for dropsonde report . . . eye oval shaped . . . estimated diameter ten miles . . . small eye . . . wind 10-15 knots . . . thunderstorms towering over forty thousand feet in the wall cloud . . . specks of foam from wind riding gigantic swells rolling toward the center of the eye from all directions.' As these swells crashed together I could see large geysers of spray rising toward us. In the midst of all of this chaos, I saw a white bird calmly floating on a chunk of debris! Call it a miracle of nature, but only a few miles from our quiet position in the eye, winds were howling up to 200 MPH!

"Aircraft commander, 'tighten up . . . we are leaving the eye.' The plane was moaning and groaning and, suddenly, . . . a giant hand was thrusting us toward the heavens. In the next instant, we began to fall. We hit the bottom of the downdraft with a shuttering bang as the pilots struggled to keep the plane under control. 'AC to crew, I want a damage report.' The front crew was shaken but no damage. From the rear, the right scanner reported, 'Sir, sir, the relief can broke loose and its all over everything!' He paused and yelled, '. . . the dropsonde operator's belt broke; he hit the ceiling bulkhead and he's out cold!' The pilot called for a heading to the nearest runway, after the flight engineer reported losing brake fluid. The navigator reported, 'Sir, we are five hundred miles from Iwo [Jima]—we can be there in two hours and twenty minutes.' 'AC to engineer, will we have any brakes left?' 'Yes sir, we could have about a fourth of the reservoir left if we can climb to a colder temperature.' 'Second navigator to AC, we are getting things cleaned up back here—the odor is terrible—dropsonde operator is conscious, but a cut on his head.' . . . 'AC to radio, tell Guam we are aborting to Iwo.' "

Captain Tucker landed at Iwo Jima using the entire runway. The duty officer told the crew that *Patsy* was headed for Iwo and everyone must evacuate! Tucker told the officer he didn't give a damn if *Patsy* was coming, he wanted the brakes repaired. "Sorry," the officer said, "all repair crews have left." McFarland and his crew were stuck for the night and this time they would "fly" *Patsy* from the ground. They were ordered to spend the night in the shelter in Mt. Suribachi while the copilot and engineer would "ride out" the storm in the plane. They would taxi the WB-29 into the wind, feather the engines, keep the plane headed into the storm with the rudder, and "fly" the plane back to the ground if it became airborne— winds of more than 120 mph can lift off a B-29.

"Sometime during the night, the wind reached a high pitched wail— like a jet engine and began to diminish. We woke a few hours later as fresh air was blowing through the shelter. . . . Our big bird was still on its hardstand—it had ridden out the storm twice! All buildings were gone except base operations; every Quonset hut had been blown away, and large refueling tankers were rolled over. At the storm's peak, the wind was over 120 MPH and the B-29 had lifted off the ground twice but had survived! We returned to Guam by C-54 as the brakes were being repaired and chalked *Patsy* up as just another routine mission."[3]

Another report on Typhoon *Patsy* comes from Leon Attarian, scanner for the 54th WRS, from Glenmont, New York. He reported his encounter: "Our crew was assigned to penetrate Typhoon *Patsy* approaching Iwo Jima. Arriving over Iwo, we saw the storm had washed over the island. . . . I could see the foundations where buildings had been. Everything on the island was gone! Personnel were saved by taking shelter in Mt. Suribachi. We contacted survivors on emergency radio and reported the situation to Japan. Within hours, a C-124 arrived with 'bare base' equipment and the base was operating in less than twenty-four hours."[4]

During the many years of weather reconnaissance, aging aircraft lacking maintenance specialists often flew in dangerous situations. Because of inoperative radar on many missions, crews made typhoon penetrations at fifteen hundred feet so they could to see the surface and be guided into the

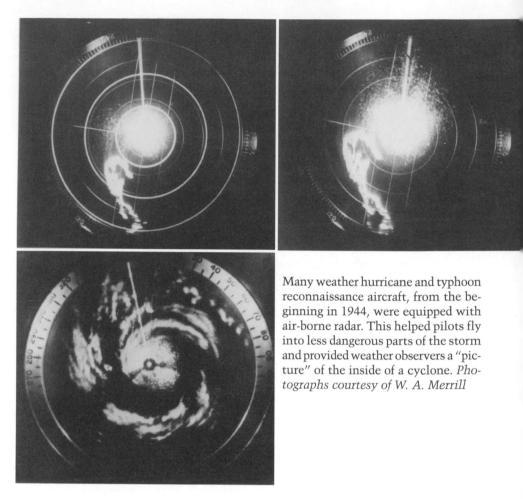

Many weather hurricane and typhoon reconnaissance aircraft, from the beginning in 1944, were equipped with air-borne radar. This helped pilots fly into less dangerous parts of the storm and provided weather observers a "picture" of the inside of a cyclone. *Photographs courtesy of W. A. Merrill*

storm by maintaining a 90-degree position to the wind. Penetration at fifteen hundred feet was more treacherous than at the normal 10,500-ft. level. Maintaining altitude was critical because as pressure changed in the storm, the pilot had to constantly keep adjusting the pressure altimeter at the direction of the weather forecaster who had the radio altimeter in the nose. On one fifteen hundred-foot flight, aircraft commander James Slaeker, Federal Way, Washington, asked the navigator the location of a small volcanic island that was about fifteen hundred feet—the plane's altitude. "Before I could get an answer, I saw the top of the island as we skimmed by with little clearance. At that point, we went to 10,500 feet in a hurry," Slaeker said.

Maintenance crews tried desperately to keep aircraft flying, often by taking engines from others that were grounded while waiting for parts. At one time in the 54th WRS, one squadron had three cannibalized planes sitting on the ramp, just to keep one in the air. Security was lax. A crew came to the flight line early one morning to pre-flight the aircraft and found initials scratched on a couple of the props. Friendly natives were intrigued by the huge planes flying around their islands, and their reactions were entertaining. For example, on one long flight from Guam, which the navigators thought "went west forever," the tiny island of Kapingsmarangi offered burlesque relief from the boredom. Leon Attarian reported that "The native women wore wrap-around skirts of some kind—mostly red. They could see us coming and as we passed over, they whipped off their skirts and waved to us. Unfortunately, one pass was all that we had time for."[5]

Samuel T. W. Davidson, Jr., a 54th WRS navigator, kept crew members entertained with his one-liner sense of humor. The relief tube of any aircraft is an important funnel-like piece of equipment with a tube to a can. Since the WB-29s were pressurized, the tubes could not empty outside—as they did on the older non-pressurized aircraft. However, in rough weather, with the aircraft bumping around, to use the relief tube was to walk away with wet hands—a constant complaint to maintenance. One day, the crew was pleased to see that maintenance had installed a handle on the funnel, allowing its use with hands at a safe distance. Davidson called, "Hey, guys! Whaddya think of the handle on the relief tube?" "Great," said a voice over the intercom, "A tremendous improvement on the quality of life up here." "Yeah," said Davidson, "Have you noticed that your sandwiches taste better now?"[6]

Because of the long missions, hemorrhoids were the most common complaint. Crew members sat on packed exposure suits and they weren't soft. Bodily functions had to be restrained for long hours in the air. In addition to the relief tube, the B-29 had a waste-paper size can for other purposes for twelve crewmen. The forward crew had to crawl thorough the tunnel over the bomb bays, and the crewmen in the rear weren't happy to have visitors come and do their business—things did not smell exactly like perfume during a long flight. Most crew members trained their bodies to "go" before the mission at 4 to 5 a.m. and wait until the end of the mission at 9 or 10 p.m.—always a hardship. If there had been a medal award for thirty missions of "Superior Control of Bodily Functions"—weather crews would have won, pants down![7]

Meals were also problems of long missions. A fourteen-hour flight required two meals for each crewman. David Magilavy, weather officer, reported, "If you ordered two box lunches, the second would be spoiled before the next meal. The most popular choice were IF2 in-flight lunches since each contained a can of meat product. Whether you got hamburger, boned chicken, roast beef, or pork and apple sauce, they were all similar in taste although the chicken was the best and the pork and apple sauce was 'yuk.' When we found someone who liked pork and apple sauce, we always selected them for the missions for food swapping purposes."[8] John Field, Issaquah, Washington, scanner for the 514th WRS, warned against eating certain foods during a storm, ". . . the wild turbulence just before entering the eye of a typhoon was precisely the time I attempted to put jelly on a cracker, ending up with jelly dripping from the visor of my cap."[9]

The Boeing WB-50 was the next aircraft to become a part of the heritage of weather reconnaissance. Boeing first designed it as the B-29D, but it was later renamed the B-50, and was first flown on 25 June 1947. The WB-50 was the same size as the WB-29 except that the 3500-HP Pratt & Whitney R-4360-35 engines required larger nacelles and a larger vertical stabilizer and rudder to offset the increased torque of the powerful engines. For this reason the WB-50 was five feet taller than the WB-29 and the rudder folded to accommodate existing hangers. At a glance, the WB-29 and the WB-50 seemed identical except for the larger tail and nacelles. Pilots liked the greater power of the WB-50, and it was exceptionally stable.[10] The weather service liked it because it added about 850 miles to the range of weather missions and completed regular missions in a shorter time. While the WB-29 looked every bit the part of the "aristocrat of the skies," the WB-50 looked tough, like a street fighter designed for the back alleys of the world's weather. It commanded respect, both from the crews that flew it and the historians who write about it.

However, John Fuller, a most respected weather service historian, calls the "saga of the WB-50 one of the blackest pages in AWS history."[11] The Air Force disagreed. A news release from the 56th WRS, at Yokota AFB, Japan, on 16 September 1965, proclaimed, "The last WB-50 took off on a trip that marked the end of a ten-year career as one of the Air Weather Service's most reliable and versatile aircraft. . . . No airplane has yet surpassed its ability to accomplish Air Weather Service's variety of missions . . . high, low, long and far, through all kinds of weather from the Arctic to the equator. . . ."[12]

With such diverse views, as a historian for weather recon, I tread lightly in describing and accessing the position of this huge aircraft that followed the WB-29. The WB-50 began service with the 55th WRS in 1955, and the 55th played an important role in the transition and training of weather recon units from WB-29s to WB-50s. Instructors from the 55th were sent to McClellan AFB, California, to train crews from other squadrons, and the 55th WRS furnished test crews to Lockheed Aircraft Service

The WB-50, converted for weather reconnaissance, was a much praised and maligned aircraft. It was faster than the WB-29 it replaced and flew ten years in reconnaissance. The WB-29 was a rapier and the WB-50 was a club—both were needed. *Photograph courtesy of Glen Sharp*

to test planes converted from B-50s into WB-50s.[13] The first WB-50 aircraft for the 56th WRS landed at Yokota AFB, Japan on 4 February 1956 and was christened "Ginkaza," or "Divine Wind," by kimono-clad Miss Hanako Shigemitsu, daughter of Japan's foreign minister. She charged the great plane to "bring benefits to all fishermen, farmers and people of all walks of life of all nations."[14]

So, why did historian John Fuller say that the WB-50 was a black page in Air Weather Service history? First, getting the new planes was a circus of military wishy-washing. When AWS wanted the WB-50, the Air Staff said it was not available since weather service did not have a high priority. When the WB-50 became available two years later, AWS did not want it. Weather was happy with its WB-29s and WB-50s were more expensive to operate. However, weather officials had no choice when they learned that twelve of the 54th WRS's WB-29s would not be able to fly typhoons during the 1955 season because of air frame corrosion. AWS was disappointed to receive yet another war-weary surplus aircraft. The WB-50 became, according to Fuller, a "widow maker." Crew transition to the WB-50 was deceptive. Lockheed had warned that the WB-50 was 75 percent changed from the WB-29, but because the two aircraft looked so nearly alike, the differences were not taken seriously. Between 1956 and 1960, there were thirteen accidents, costing the lives of sixty-six crewmen during "the blackest pages in the history of weather recon."[15]

David Magilavy, as a weather aircraft research officer at Wright Field, Dayton, Ohio, remembered that on one routine test flight, the B-50 just flew into the ground at a steep angle. Several other losses were of the same nature, and the number of crashes raised concern at the National Weather Service. George Thurman, Jenks, Oklahoma, pilot and maintenance officer, often spoke of the difficulty of maintaining weather aircraft because they were usually left over from tactical or strategic functions and parts were difficult to get.[16] The WB-50 had so many problems, it was often considered an aeronautical "Peter Principle,"—a WB-29 engineered above its ability to perform.

Aircraft were often "loaded" with excessive equipment and refinements not needed in the theater or the duty to which they were assigned, and field modification was as important as field maintenance for the WB-50. John D. Gilchrist, Lutz, Florida, maintenance engineer and scanner for the 58th, 55th, and 53rd WRSs, explained some problems of the W-50 and how they were solved in the field: "Although the WB-50 was a decent aircraft, converting from the WB-29 to the WB-50 wasn't all smooth. For example, pressure pumps for the magnetos, for high altitude flying, were not reliable, so we removed them. . . . The exhaust ports were often improperly fitted into the cylinder housing and had a tendency to work loose. . . . The first R-4360-35 engines had external oil tubes on the lower half of the engines, and they would work loose or break. On engine maintenance and updating, they were done away with. . . . The engine oil breather system had a regulator to maintain positive crankcase pressure. These regulators would freeze, and pressure would build up and blow oil out of the engine. They were removed because they were not effective. Oil consumption charts and the engine analyzer system were good to help pin-point problems, and, with field changes, our engines' life actually increased. Untrained personnel were our main problem—men who had never worked on 29s, 50s, or their engines. . . . While I was assigned to AWS, we lost two WB-50s. One plunged in from eighteen thousand feet, attributed to auto-pilot malfunction. The second was lost on take off when all four engines quit because of carburetor ice."[17]

Leaky fuel cells grounded the entire WB-50 fleet in May 1960, and the venerable and aging aircraft served their last year in 1965. Despite troubles, the plane had served well. Most made their long last flight from far-flung bases to the aircraft graveyard at Davis-Monthan AFB, in the desert of Arizona. The last WB-50, No. 49-310, was flown to Davis-Monthan to be processed for display at the Smithsonian Air and Space Museum.[18]

Pressure was a standard measure of altitude on weather flights, and missions were planned at millibar (pressure) levels rather than a certain number of feet above sea level. A millibar (mb) is an line of equal pressure, and as this line dips and rises, forecasters have a valuable tool in predicting weather movement. Normal sea-level pressure is 29.92 inches of mercury or 1,081 millibars. As altitude increases, pressure decreases. Pressure variations are extremely important to pilots, especially on long over-water flights. As aircraft move through changing pressure areas, the flight altitude above sea level changes. Pressure charts were plotted for millibar/feet levels:

mb	feet
1,081	sea level
850	4,780
700	9,880
500	18,280
400	23,570
300	30,070
200	39,000[19]

Weather flights were usually flown at the 300-mb level or lower, depending on the mission, weather conditions and the aircraft. However, ambition or the desire to explore new frontiers often created missions beyond the ability of aircraft and crew. For example, Lt. Col. Paul A. Steves, 57th WRS weather recon officer from Woodbine, Maryland, in late 1956, flew a WB-50 on the first 200-mb mission ever attempted in a propeller-driven plane. Summing up the mission, Colonel Steves, wrote, "I could write a book about what went wrong. It was a near disaster. Our 57th WRS normally flew the *Petrel Foxtrot* mission north out of Hickam AFB, Hawaii, at fifteen hundred feet, to burn off fuel, then climbed to the 700-mb level (9,880 feet) heading northwest to the Gulf of Alaska until we were light enough to climb to the 500-mb level (18,280 ft.), where we would turn southwestward and head back to Hickam. Since the WB-50 was supposedly a much better airplane than the WB-29, the 200-mb mission was planned to see if we could climb higher and make a return at thirty-nine thousand feet.

"A select crew was put together, with senior members in every position. The squadron senior pilot was in the left seat. We droned along in boredom and apprehension as the pilot pushed the WB-50 to higher altitudes. All the time we were above a continuous undercast and were not able to make 'double drifts' for accurate wind data. When we reached the 200-mb level, I noticed that the pressure pattern I was developing did not fit the navigator's report of our position. The aircraft commander asked the second navigator, a young lieutenant, to check our position. He found that the first navigator had compensated for the jet stream from the east rather than from the west, as it actually existed. The second navigator shot sun lines and announced that we were nowhere near where we thought we were, and, worse yet, we were nowhere near land we could reach with the fuel we had remaining.

"Troubles mounted. No. 4 engine developed internal problems and the engineer announced that he was feathering it. The aircraft commander asked for a heading back to Hickam where the water was warmer if he had to ditch, and he was going to make the longest glide in Air Force history! We called Hickam that we would ditch two hundred miles out and to send Air-Sea Rescue to that spot. With real piloting skill, navigational accuracy,—the young lieutenant was good—a little pressure-pattern flying, and luck, we passed the rescue C-54 at the right spot and made it home with two engines quitting from fuel starvation on the flare-out for landing. In just a few years, jet aircraft made higher altitude flying routine, but at that time, getting a tired old prop plane to perform a 200-mb weather mission was not easy. That was a harebrained mission!"[20]

At the next day's briefing to the commanding general of the Air Weather Service, Steves made a few adverse comments about the "harebrained mission." That night at a cocktail party, Gen. Thomas Moorman advised him that the "harebrained mission" was his idea. "Whammo! . . . my career was doomed!" Steves wrote. "However, before I was due to return to the States, a direct message from General Moorman ordered me to Wright-Patterson AFB to set up and operate a flight test program for development of airborne weather reconnaissance equipment!"[21]

Project *Stormfury* was a hurricane modification program that was much more ambitious than mere cloud seeding. *Stormfury* had been first proposed in 1961 by Dr. Robert H. Simpson, director of the National Hurricane Research Laboratory, in Miami, Florida. Dr. Simpson's idea, based on experience in hurricanes *Esther* and *Beulah*, was that hurricanes might be modified by the introduction of a "freezing nuclei" into the clouds swirling around the center of the storm to dissipate its strength. The Weather Bureau, using the 53rd WRS and the Navy, experimented on Hurricane *Esther* in September 1961 with a dry ice seeding—one seeding on each of two days. Also, in August 1963, Hurricane *Beulah* was seeded once a day for two days. The results were encouraging but inconclusive.[22] Coincident to this idea, the Navy Weapons Center, China Lake, California, developed pyrotechnic generators capable of dropping large quantities of silver iodide crystals into storm clouds.

William Anderson, Boise, Idaho, typhoon recon pilot of the 57th WRS, gives his evaluation of the hurricane seeding efforts: "*Stormfury* was a . . . , a kind of atmospheric judo—a karate chop to the elements. Man cannot create anything approaching the energy of a hurricane, according to meteorologists. Even an H-bomb exploded in its core would have the effect of a popcorn fart. *Stormfury* attempted to use the giant's own energy against itself, setting in motion a chain of reactions within the storm that would finally dissipate its energy."[23] Weather modification has always had its utopian enthusiasts who believed that changing weather was possible and the key to the good life, while dooms-day prophets warned of "weather going wild" because of man's tinkering with Mother Nature. Even During World War I, when Europe and parts of the United States had an extremely wet season, the rains were blamed on artillery fire on the western front.[24] President Kennedy, on 22 October 1963, remarked in a speech to scientists, "We must balance the gains of weather modification against the hazards of protracted drought or storm."[25]

It was only natural that some attempt would be made to modify killer hurricanes, the most destructive of all storms. As lives and property from the Florida Keys to Maine, and along the Gulf coastline to Mexico, were threatened and destroyed year after year, meteorologists and ordinary citizens, asked many times: "With modern technology, why is it not possible to kill hurricanes before they become a threat?" This question is one that has intrigued weather scientists through the centuries. There have been many ideas to stop or change the courses of hurricanes while they are still well out to sea, from dropping atomic bombs into the storm, to cloud seeding. The atomic bomb has perhaps made man feel invincible, but few people can imagine the raw power in these killer storms. The atomic bomb is puny compared to a hurricane, which generates more energy than a dozen Hiroshima-size atomic explosions.[26] Attempts have been made to destroy storms, but not with atomic bombs because of the released radiation. The best idea seems to try to turn the energy of the storm onto itself, such as an early attempt at cloud modification, from 23 to 31 July 1958, when the Navy Weather Service conducted tests to destroy clouds with the use of air-dropped carbon black.[27]

Seeding hurricanes seemed to be the most practical idea. However, remembering an earlier hurricane seeding incident, the project directors moved slowly as they began their quest for just the right storm. A hurricane in 1947, before storms had names, had already passed over Florida and, after seeding by the Navy, "turned inland and clobbered Savannah,

Georgia,"[28] then ran wild and turned back across the coasts of Georgia and South Carolina in what seemed to be atmospheric revenge. This seeding brought a political storm worse than the hurricane. Many people believed that seeding the hurricane had caused it to change course, threatening their lives and property. This downside in all weather modification experiments produces a paranoia that asks many questions: "Did the storm threaten my life and tear up my property because it was seeded?" or "Did the rain that was scheduled for my farm fall on my neighbor's?" or "Would it have been the same without the experiment?" No one can prove any of these points either way. James F. O'Connor, expert of the Weather Bureau's long range forecast team, points out, ". . . until weather can be predicted precisely, scientists will always be doubtful of weather control claims. They will wonder if the changes would have occurred anyway."[29] As a result of the 1947 incident, policy was adopted that seeding could only be done on storms with a one-in-ten chance of hitting a populated area within twenty-four hours after seeding. [30]

Stormfury was a joint project with the Department of Defense, using Navy and Air Force weather crews, and the Department of Commerce's Environmental Science Services Administration (ESSA). In August 1969, *Camille* and *Debbie* were two storms that became candidates for experiments, and *Debbie* was selected as being the safer of the two. *Debbie* was assaulted by two hundred men and thirteen aircraft, five of which were fitted with special wing racks to hold silver iodide canisters. The other aircraft were equipped to measure temperature, wind speed, and cloud moisture before and after seeding.[31] Leslie Nelson, Sierra Vista, Arizona, flying out of Ramey AFB, Puerto Rico, was a dropsonde operator for the 53rd WRS on the *Debbie* missions. Nelson remembers, "Our mission was to dissipate or modify tropical storms. . . . The Navy did seeding at 500mb (18,280 feet) and the Air Force, flying WC-130s, entered the storm at 300mb (30,000 feet) to record temperature data, wind, and cloud changes. The 53rd crews' main mission was to observe and measure changes in the eyewall or rain bands caused by Navy seeding. Cecil Gentry, director of the *Stormfury* project at that time, was enthusiastic. He called the *Debbie* experiments an "operational success" in that they were able to put the silver iodide where they wanted it; it was the type of storm they wanted and the aircraft collected the type of data needed. But the scientific success was uncertain, i.e., no one could be sure how the seeding affected the storm's intensity or movement. Gentry added, "Whether prompted by man or following its own energies, *Debbie* turned sharply away from land . . . and headed for the open sea." The *Stormfury* director claimed no responsibility for the change of course, but added, "If the hurricane had hit land, we would have had a lot of explaining to do."[32]

On June 6, 1971, in a state-side project, the 55th WRS seeded clouds in

Southern Texas in an attempt to alleviate a serious drought. The missions were so successful that a tactical airlift group from New Mexico was called in later to help evacuate flood victims. The 53rd flew inconclusive seeding missions in Hurricane *Ginger*, on 26 September before the phase-out of the experiments.[33] The *Stormfury* missions were somewhat successful, but were ended because of political ramifications of modifying storms over international waters.[34] A plan, in 1972, to move *Stormfury* to the Pacific was abandoned because of opposition from China and Japan.

Although they were called *"Hurricane Hunters,"* crews of the 53rd WRS were busy flying all over the world with other than routine assignments. They were the first to fly over Iceland to collect air samples above the Mount Hekla volcano as a project of the National Center for Atmospheric Research. They then went to Hawaii to support the downrange recovery of space capsules. Then the 53rd flew to Europe where they became "fog floggers" to clear the winter skies,[35] and in 1971, as hurricanes and typhoons went wild in the Atlantic and Pacific, the squadron was kept busy flying storm missions. For the first time, the 53rd was called to the Pacific to help the 54th WRS *"Typhoon Chasers"* with typhoon recon. In July, they were in the Western Pacific in support of the Apollo 15 launch. Back in the Atlantic, in September, there were five named hurricanes prowling the sea lanes at the same time, and the 53rd flew 415 hours in forty-four sorties. A storm named *Ginger* roamed around the Atlantic for thirty-one days and became the longest lived hurricane on record.[36]

The Cuban missile crisis, in 1962, brought the might and strength of the U.S. into action. With the nation's attention focused on the Vietnam war, the USSR began building medium- and intermediate-range Il-28 missile launching sites in Cuba. President Kennedy quickly flexed the nation's military muscle against Fidel Castro, his "bearded and blustery buddy in Havana, the gadfly of the Caribbean."[37]Cuba was cut off from the world by a Naval blockade, military reserves were called to duty, and the island became a potential target for every weapon the U.S. had. Guns were being cocked throughout the Western Hemisphere. Kennedy demanded, on 22 October, the withdrawal of missiles and on 28 October, the Soviets agreed to dismantle the launching platforms. The Strategic Air Command (SAC) asked Air Weather Service to make reconnaissance flights around Cuba. For two months, beginning on 30 October 1962, the 53rd's weather-weary WB-50s flew a daily circular, 3,857-mile track from Bermuda around Cuba, carefully staying forty miles from shore, and returning to Bermuda—a sixteen-hour flight. These missions, operating under the code name *"Easy Aces,"* were under the direction of the AWS's commander Norm Peterson.[38]

In 1963, with the Air Force moving toward an all-jet fleet, weather reconnaissance was still flying propeller-driven planes, limited in speed and altitude. Storms prowling the vast distances of the Pacific sometimes moved faster than the WB-50s, and aircraft often had to fly for hours before reaching the storm area. New assignments for weather, such as support for space probe launch and splashdown, demanded faster, high altitude aircraft. As the aging WB-50 fleet was being mustered out, the WB-47E, a modification of the Boeing B-47 *Stratojet*, the Air Force's first strategic jet bomber, introduced the jet age to weather reconnaissance. Lt. Col. John D. Horn, Whitefish, Montana, commander of Detachment 2 of the 55th WRS, was enthusiastic: "The WB-47E will do a better job, at a higher altitude and in half the time of the WB-50."[39] There seemed to be a bit of status in the adoption of the WB-47E since the AWS was the last of the Air Force divisions to use jet aircraft.

The first production model of the WB-47E was flown from McClellan AFB, California, to Hickam AFB, Hawaii, in five hours and forty minutes, and the first of thirty-four WB-47Es was delivered on 20 March 1963. The first weather mission was flown on 6 July 1963.[40] These swept-wing super jets were equipped with the AN/AMQ-19 meteorological system, the latest in weather observing systems.[41] WB-47Es were modified by the Lockheed Aircraft Company at Marietta, Georgia, and the changes included removal of all bombing systems, installation of in-flight refueling capability, and installation of the latest weather equipment.

Despite its thin, sleek look, the WB-47E was a huge aircraft, almost as large as the WB-50, with a wingspan of 116 feet, 25 feet less than the 50's 141-foot span. However, the 47 was eight feet longer at 107 feet. The knife-thin, swept-back wing had two hundred square feet less area than the WB-50; the two aircraft were the same weight empty, but the WB-47E could carry thirty thousand pounds more gross weight. To move this huge plane through the air at 500-MPH (top speed, 600-MPH) up to forty thousand feet required the power of six General Electric J47-GE-25 jet engines, each with seventy-two hundred pounds of thrust. On each side, two engines hung in a dual pod below the wing near the fuselage, and one engine was near the wing tip. The 47 had a tandem landing gear with two small outrigger landing wheels and a brake 'chute for landings. Comparing the WB-47E to the WB-50 was like comparing the rapier to the club although many times weather recon needed a club.

In the seemingly random movement of weather recon squadrons from base to base, the 53rd was moved to Hunter Air Force Base, Savannah, Georgia, on 8 September 1963. For the first time in seventeen years, the *Hurricane Hunters* were in the United States, closer to the hurricane

Weather reconnaissance received a new super-sonic aircraft when the WB-57 joined the fleet. The 57 dictated a new type of recon -- it flew faster and higher than any other weather aircraft. *Photograph courtesy of Eugene Wernette*

breeding areas and coastal storm paths as the cyclones roamed the Atlantic, Caribbean and Gulf of Mexico. At Hunter, weather recon had entered the jet age with the addition of WB-47Es to the fleet. The first WB-47 was christened *"The City of Savannah,"* and one newspaper reported a significant historical fact: the commander at Hunter, Lt. Col. Arnold Zimmerman, Sacramento, California, happened to have the same air force serial number as one of the first WB-47Es to fly from Georgia. Although the squadron was officially back in the States, Detachment 1 of the 53rd, remained at Kindley and continued to fly WB-50 recon from its Bermuda base.[42]

The all-jet super-sonic weather fleet at Hunter consisted of nine WB-47Es and twelve crews, with the first aircraft delivered on 17 September 1963. The first hurricane mission for the 47E was flown into Hurricane *Cleo* on 28 August 1964, led by Lt. Col. Eugene Wernette, aircraft commander; Maj. Vernon Blankenship, pilot; and Capt. James Glover, navigator.[43] With the new jet aircraft, the techniques of hurricane surveillance changed drastically. The older WB-29s and WB-50s carried a crew of eight or ten on missions from fourteen to sixteen hours at 300 MPH. Storms were penetrated twice or more, and, before the mission was complete, crews found the eye when possible. The WB-47E, a mach-speed, very high altitude aircraft, carried a crew of three and could complete a mission twice as fast—at 600 MPH instead of 300. Whether the WB-47 aircraft was an improvement over the older WB-29s and WB-50s is debatable. Because of its great speed, the 47 could not penetrate the interior of storms because the aircraft was not stressed for hurricane flight, and extreme turbulence could rip the wings off. The speed of an aircraft is a factor in its ability to fly into turbulence safely.[44] Consider this comparison: Driving your Mercedes down a very bumpy highway at 60 MPH rather than 30 MPH would give the world's roughest ride, severely damage or rip off the undercarriage and, possibly, wrest your car out of control.

So WB-47Es were not flown into hurricanes, but their high speed enabled them to reach the storm area more quickly, to skirt the edges, circle the storm and, possibly, fly over the top at thirty- to forty-thousand feet. Above the swirling hurricane, a dropsonde-type sensing instrument was released and as it fell through the storm, it radioed reports of temperature, humidity and pressure until it crashed into the ocean. The new jet fleet did not solve all recon problems, and there was always a question of its being able to record enough useful information simply by circumnavigating or flying over the storm. So, in what must have been slightly embarrassing to modern jet technology, the 53rd called in three tough old WB-50s from the 56th WRS in Yokota, Japan, for temporary duty to help do the job the WB-47E could not do. A Savannah newspaper proclaimed, "A seasoned veteran of the storm wars today [16 June] joined nine brash and swift WB-47E newcomers at Hunter AFB in the job of hunting the gales that blow up into hurricanes. . . . The 'old pro' propeller-driven WB-50 had been chasing typhoons in the vast Pacific and flew in from Yokota AFB, Japan, under the command of Lt. Ben T. Furuta and a crew of nine. In tracking hurricanes, the WB-50 can fly into the storms at ten thousand feet, something the jet-powered WB-47E dared not do."[45] The WB-50s were able to penetrate at lower levels, take internal observations in the storm and measure the swirling winds. However, with both WB-50s and WB-47s, crews were able to get a double fix, providing more information that ever before. The Navy was responsible for reconnaissance east of the Winward Islands and for night recon because of its efficient radar.[46]

The new weather jet carried a crew of a three—aircraft commander, pilot and observer— sitting tandem in a pressurized cabin, and the crewmen wore pressurized "space-type" flying suits. With no way to move around to stretch their muscles, pilots noted that on a long flight their butts and legs would become numb. David Magilavy reported, "Being of a scientific bent,

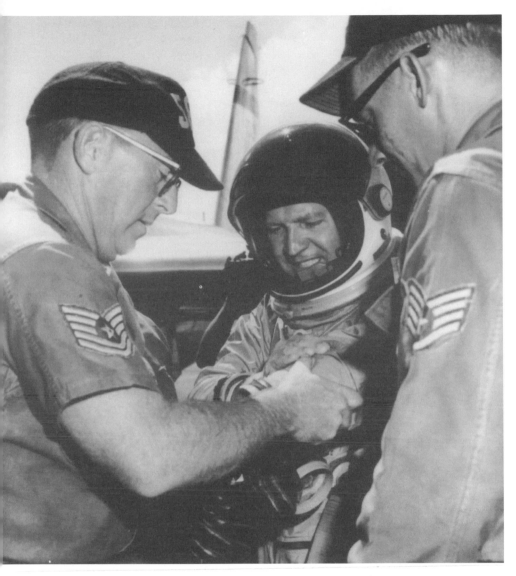

Robert Moeller, aircraft commander of a WB-57 is helped into his "space suit" in preparation for a flight. The WB-57 carried a crew of three and could locate and scout storms faster than any other aircraft. However, it could not safely penetrate the deadly storms. *Photograph courtesy of Robert Moeller*

one pilot made time studies and found that he had a three-hour butt and three and one half hour legs."[47] Tandem seating was a minus in planning because the cockpit was "designed for fighter pilots." Limited visibility in the rear seat and, with only engine RPM gauges and poor flight instruments in the rear cockpit, instrument flying, landings or emergencies requiring pilot coordination were difficult.

In weather recon, the inability of the WB-47E to penetrate a hurricane was a concern. Although the 47 was to be used for hurricane and typhoon reconnaissance, Boeing had recommended against it. The outer tips of the wings flexed up to 21 feet in severe turbulence, and the plane was not designed for such stresses. Jet experts were also worried about a possible flameout in heavy rain. So, with every mission of the WB-47E, there was anxiety about its ability to stand the rigors of normal weather flight. Actually, it was suspected that the aircraft had been selected for air sampling rather than normal weather recon.[48] The first WB-47E fatality was a crash on landing at Lajes Field, Azores, on 23 November 1963, and a second crash came five months later at Eielson Field, Alaska.[49]

The WB-47E aircraft had limited ability for weather observing since it was not able to carry the specialized crew or instruments of the older planes. In an attempt to supplement or replace human weather observers, the AWS had tried for many years to develop a system of automatic observation and data transmission as the aircraft penetrated storms and other weather—a sort of horizontal radiosonde. Thus began a series of "AN/AMQ" systems, a "black box" recorder-transmitter of various configurations, depending on the model. The systems recorded horizontal data and, coupled with the AMT dropsonde systems, recording vertical data, it seemed that automatic, un-manned, weather observation had become a reality. Some systems were successful, and some were not, and research was canceled many times at the whim of the military counting house.

Clyde Garrison, SAC B-47E and AWS WB-47E aircraft commander with over four thousand hours of 47 flight time, gives the plane a mixed rating for weather. The 47 was a high speed (.74 mach); high altitude (37,000 to 45,000, depending on weight), and there was a surplus of good SAC pilots with B-47E experience. However, the plane had high fuel consumption, and maintenance was a problem when operated globally because of a shortage of skilled maintenance personnel. And the B-47Es had ten to fifteen tough years on them before they were modified for weather.

One dangerous characteristic of the WB-47 was pointed out by Garrison: ". . . with the loss of any one of the inboard engines, (Nos. 2, 3, 4, or 5), after max refusal speed on normal take off, flap retraction, and climb speed, there was no control problem and plenty of power. However, if you lost one of the outboard engines (Nos. 1 or 6) . . . because of the swept wing, the pilot had to immediately give three-quarters to full opposite rudder, or, within three to five seconds, the plane could stall out of control. This problem was not known until 1962 by either SAC or Boeing."[50] Garrison felt that the weather service was lucky in that they didn't keep the WB-47 in its inventory long enough for the "law of averages" to catch up.

Thomas Robison, atmospheric equipment repairman from Ossian, Indiana, remembers well when the 47s were being phased out. "No tears were shed," he said; "The Air Force and the Ninth Weather Wing were anxious to get rid of the WB-47. Actually, not long before the 47s were being retired, one plane had a belly-in due to gear failure. It was piloted by the Ninth Wing safety officer. The damaged belly panels were dinged out, outboard engines were replaced, and it was flown to Davis-Monthan, the aircraft graveyard, without a check flight."[51] On 2 September 1969, Air Force Chief of Staff John D. Ryan ordered all twenty-four remaining WB-47Es to "depart for Arizona," and the 57th WRS, at Hickam, Hawaii, deactivated. The 57th WRS was officially deactivated on 30 November 1969, and the remaining WB-47Es took their last flight to aircraft Valhalla.[52]

Another weather jet aircraft, the WB-57F and its high-altitude version, the RB-57, expanded weather reconnaissance to near-space altitudes. These two aircraft were an extreme modification of the Martin B-57A British *Canberra* with a totally different wing and power plant. The weather version was modified from older B-57 models by General Dynamics of Fort Worth. It was powered by two Merlin Rolls-Royce engines that were started by a black-powder charge. Two TF-33-P11 turbofans and two J-60-P9 auxiliary jets in underslung pods were installed on a new 122-ft. wing, almost double the original 64-ft. wing of the B and C models, and drooped like the WB-47E. The WB-57F was used in atmospheric sampling missions by the 58th WRS at Kirtland AFB and operated at near U-2 altitudes, flying weather missions up to forty thousand feet. The RB version could take long range sampling missions of radiological ash above sixty thousand feet. The two-man crew wore $7,000 pressure suits and were transported to the already preflighted aircraft in an air-conditioned van. Treated like astronauts, the two men got in the 57 and took off. The WB-57F was also used by the 56th WRS in Yokota, Japan for classified missions.[53]

Crews for the RB-57 version were treated like astronauts, and during the seventy-two hour period before each flight, the two-man crew ate specially prepared food and maintained rigid exercise and rest rules to insure peak condition. A flight physical was given before and after each flight. Donald Schertz, Monterey, California, did testing on the suits. He explained that "the pressurized flight suits used physical pressure to counter the oxygen pressure within the suit, and although they had been checked

The Martin RB-57A, used in weather recon for a short time, was a converted version of the British *Canberra*, and was capable of almost U-2 altitudes. Crews wore space suits and went through a rigorous pre-flight regimen. *Photograph courtesy of John Pavone*

in decompression tank tests, the crews' skin revealed hundreds of tiny broken blood vessels, looking like sunburn or the measles."[54] Crews also lost several pounds of weight during each flight.

The first of nineteen RB-57s was delivered to the 58th WRS, Kirtland AFB, New Mexico, on 18 June 1964, and the first WB-57F was delivered from Kirtland to Eielson, Alaska on 30 October 1964, to replace WU-2 aircraft. Flights were under the direction of Lt. Col. Robert L. Moeller, squadron commander.[55] A series of crashes at Kirtland, two within a twenty-four hour period, emphasized the difficulty of flying the hot new aircraft. One accident was caused by a flame-out, and the aircraft landed on a mesa five miles from Kirtland; a second crash landed on the runway and burned. Crew members escaped without injury although two crew members were killed in a 57 crash in the Sandia Mountains, near Kirtland, on 7 November 1966.[56] Most crashes were attributed to engine failure and loss of control. A full-power rudder boost was added to the planes, but improper engine operation techniques remained the cause of accidents. The minimum air speed for control was 155 knots (178 MPH), but below that speed, the pilot lost control. The warning "155 IAS" (indicated air speed), several inches high, was posted on the dash to implant the figure in the pilot's mind.[57]

The WB-57Fs became increasingly hard to maintain—wing cracks were found more often—and, on 17 September 1973, thirteen of the 58th WRS's 57s were ordered into mothballs and the high altitude sampling mission was transferred to the Strategic Air Command (SAC). Paul Steves, in the Wright-Patterson AFB meteorological test program, considered the WB-57 to be the best of WB-50, WB-47E, F100F and T-33 aircraft. After phase-out in weather, no other agency was interested in this aircraft. He wrote, "I flew one WB-57, equipped with the latest AN/AMQ-15 weather system and the AMT-6D dropsonde system to the Miami Hurricane Center, in Florida, to try to interest them in using the aircraft for research. However, its limited range and rigid structure made it unsuitable for hurricane penetration, and I suspect the operational cost would have wrecked their budget." This flying weather station was later scrapped for parts.[58] In 1984, with the developing satellite program, the WB-57s were released for salvage, and the worldwide recon fleet was becoming history. Several of the aircraft were requested for static display at museums, air bases, etc.[59]

It is difficult to tell exactly just when war and the might of the United States exploded in Vietnam. The French gave up in May 1954, and the U.S. began serving as advisors to South Vietnam's government. Slowly, after back-alley guerrilla warfare, the conflict escalated, until August 1964, when two U.S. destroyers, the *Maddox* and the *Turner Joy* were attacked by the North Vietnamese. Congress passed a resolution allowing President Lyndon Johnson to invoke any measure he saw fit to prevent further attacks against U.S. forces. Very quickly, the conflict became what would be the most unpopular war in history. Events in Vietnam are vague, especially weather reconnaissance and, this seems to have been reflected in the sentiments of Adm. Thomas H. Moorer, chairman of the Joint Chiefs of Staff, in an address to Congress in 1973. Admiral Moorer said, "I do not think any useful purpose would be served by critiquing what happened in the past [in Vietnam]."[60] The same feeling was evident in research into this weather reconnaissance history. Weather recon had a reduced role in Vietnam, and of the hundreds of weather recon veterans, in all parts of the world, who contributed their experiences to this history, only one mentioned Vietnam—a short description of the SAC *Arc Light* missions. Vietnam was not a popular war.

Fighting the North Vietnamese in Vietnam was like going into the water to fight a shark. As U.S. military might fought along the jungle trails, three weather recon squadrons, the 54th, at Anderson, Guam; the 57th, Hickam, Hawaii; and the 56th at Yokota, Japan, continued normal flight schedules. Their most useful contribution would be regular typhoon and synoptic recon missions covering the area that affected Southeast Asia, and weather crews supported route recon for fighter aircraft being ferried to Vietnam. And in weather recon's decreased role came another strong suggestion that technology was forecasting the end of aircraft reconnaissance. Increased efficiency of weather radar and the use of satellites decreased the need for manned aircraft recon although target reconnaissance was still demanded by some commanders. As in the Korean conflict, there was concern about sending unarmed weather aircraft over target areas. Commanders did not want to risk their fighters for protection of the weather planes, and some commanders simply did not like to see what they considered senseless use of aircraft on nonaggressive missions. Brig. Gen. George Simler, Second Air Division, told Col. Alexander Kouts, "I am wasting six perfectly good fighter bombers on your goddam weather reconnaissance."[61]

The Strategic Air Command (SAC) usually got its way in war strategy and was one of the most enthusiastic users of weather reconnaissance. SAC's Guam-based B-52 strikes against Vietnam were called "*Arc Light*," and weather missions were requested to scout refueling areas. On 7 August 1965, supporting *Arc Light* bombing missions, Detachment 2 of the 9th Weather Wing, with WB-47Es from the 56th and 57th WRSs, began flying weather scout missions from Clark AFB, Philippines. These flights continued until the 47s were retired in 1969.[62] Three WB-47Es of the 56th at Yokota, were sent to Clark AFB in the Philippines to support KC-135 tankers refueling the B-52s on their return to Clark.[63] Later SAC requested two *Arc Light* sorties, plus one weather route mission each day. To meet this request, more aircraft were brought into Clark until the base was so crowded that Air Force Chief of Staff, Gen. John "Jack" D. Ryan, who was not a believer in weather recon, requested that weather planes leave Clark. It was claimed that the WB-47s were tearing up the runways, and personnel were taking up needed living space. However, the influence of SAC kept the weather craft flying out of the Philippines, logging more than eight thousand hours of flight time in two years.[64]

Dates overlapped as planes came in and out of the weather service. As the WB-50s and WB-47Es were being phased out, without fanfare, another propeller-driven, homely, clumsy-looking goose was moving into the weather fleet. It was the weather version of the Lockheed C-130 "*Hercules*," with a proud record in transport and battle. "A fine reliable old bird," according to

The Lockheed WC-130 *Hercules,* a four-engine, turboprop converted cargo plane, has become the most reliable of all weather recon aircraft. The *Herky* has served weather for forty years and is still on call from Keesler AFB for hurricane missions. *Photograph courtesy of Eugene Wernette*

weather pilot William Anderson, Boise, Idaho, author of *Hurricane Hunters.*[65] The WC-130 would become the most dependable and durable aircraft to fly weather.[66] Crews affectionately referred to their plane as "*Herky,*" and called themselves "*Herkynuts.*" The *Herky* was the only aircraft to have an international fan club, and amateur historians keep records to show the current location of every WB-130 ever built. (Example: WC-130A, Lockheed Production No. 3127, USAF No. 56-0518, is now "rotting in Vietnam," a derelict at Tan Son Nhut, suffering severely from neglect.[67]) The WC-130, in A, B, E and H models, served weather for forty years and today, when the weather jets have "died" and the glamorous WWII bombers are in museums, the WC-130 *Herky* is still on call for hurricane and cold weather missions out of Keesler AFB, Biloxi, Mississippi.

The WC-130 is an all-metal, four engine, high-wing aircraft with tricycle landing gear that squats, unglamorously, to bring the cargo door to ground level. Four Allison T-56 A-15 turboprop engines drive Hamilton Standard full-feathering, reversible-pitch propellers at over 1000 RPM. The plane has a wingspan of 132 feet, 7 inches, a length of 97 feet, 9 inches and, and its huge vertical stabilizer is 38 feet, 3 inches high. It has a cruising speed of 350 MPH and a ceiling of thirty-three thousand feet, carrying a basic crew of six, with plenty of room for leg stretching, instruments, radio equipment, etc. There were only forty-two WC-130s: three As, seventeen Bs, six Es and sixteen H models in weather service, and only two weather models have been lost. On 12 October 1974, No. 65-0965 of the 54th WRS disappeared while scouting Typhoon *Bess* in the South China Sea. No trace of aircraft or crew was found.[68] No. 62-3494, sold to Pakistan in the mid '80s, crashed on 17 August 1988, while carrying the president and senior military staff of Pakistan.

The first of five new WC-130Bs, fitted for atmospheric sampling, was delivered to the 55th WRS on 22 October 1962. The others followed within a month, and the plane immediately began recon missions. In the Atlantic, the first storm mission for the WC-130 was in Hurricane *Betsy*, 27 August 1965, flown by the 53rd WRS. *Betsy*, at that time, was the most destructive hurricane to strike the nation's coastline, causing $1.5 billion in damage. Pilots of the 53rd say that the WC-130 was the first airplane that could penetrate a hurricane's torrential rains without getting wet inside— unbelievable to crews who always came home from a mission soaked to their parachutes. But even if the plane is dry inside, CM/Sgt Lee R. Dunn, maintenance specialist of the 53rd, sees rain erosion, on any aircraft, as the major problem with hurricane recon. "Rain, when it is thrown by an 80- or 90-knot wind, can chew right through deicing boots on a propeller or wing. Rain wrecks a radome. We can't use plastic or fiberglass parts, and one mission is the most we can hope for with that sort of equipment." Turbulence also causes problems. Dunn says, "Now and then a rivet pops in the aluminum skin and we discovered hairline cracks in wing spars during inspections—but nothing like the rain damage."[69]

The WC-130A was "the rainmaker" in Vietnam. In June 1967, three WC-130As, painted in Southeast Asia camouflage and flying as transports, took over secret rainmaking missions, called Operation *Popeye, Motorpool*, and *Intermediary-Compatriot*, over Laos, North Vietnam, and the A Shau Valley. Between January and May 1968, the rainmakers flew 175 missions, seeding clouds with dry-ice pellets.

If the Vietnam war was unpopular, the secret rain-making weather missions over Vietnam were more unpopular and a political storm hit Washington when word leaked that weather modification was being used as a weapon of war. Everything had been top secret, and official Washington had refused to discuss rumors. A diplomatic crisis developed when *Washington Post* columnist Jack Anderson, on 18 March 1971, wrote, in part, "The hush-hush project . . . '*Intermediary-Compatriot*,' was started in 1967 . . . [and]

In the WB-29 on storm patrol, the weather observer sits in the nose of the aircraft and directs the flight into a typhoon. Here David Magilavy gives directions to the aircraft commander prior to entering the storm. *Photograph courtesy of David Magilavy*

rainmaking missions, believed to have increased precipitation over the jungle roadways during the wet season, . . . assertedly have caused flooding along the trails. . . . Only those with top security clearance knew that nature would be assisted by the U.S. Air Force. Again, the defense by military officials was that no one could prove that the rainmaking experiments were successful. Joanne Simpson, cloud modification expert at the National Oceanographic and Atmospheric Administration (NOAA), said, 'I would be grieved to see my work used for military purposes. . . . This kind of work [is to] do useful things, not destructive things.'"[70] The U.S. Department of Defense, after first refusing comment, finally admitted that cloud seeding had been used in Vietnam, probably creating distrust of federal agencies interested in weather modification.[71]

In war strategy there has always been interest in using or controlling nature as a possible military weapon. Military commanders have fought behind smoke screens, fighters have attacked by flying out of the sun, and aircraft with radar have bombed through clouds. Rainmaking is little different. However, because of local and international concern, the U.S. Congress, in 1974, passed legislation prohibiting the use of weather modification and other forms of geophysical warfare, and the United Nations General Assembly approved an international resolution prohibiting the hostile use of environmental modification.[72] This agreement was signed by the United States and the U.S.S.R., and thirty other countries in Geneva in May 1977.[73]

When the rainmaking programs were stopped, the WC-130s were sent to India for synoptic recon over the Bay of Bengal.[74] In 1969, the Philippines were suffering from a major drought. WC-130s flew seventy-six cloud seeding missions, from 29 April through 20 June, with extremely successful results. The missions were marred by a dry ice ejector explosion during flight on one of the aircraft. Weather officer Capt. Charles D. Booker was seriously injured and he died two months later.[75]

As large as the WC-130 was, pilots flew it like a fighter. Tom Robison, Ossian, Indiana, maintenance specialist for the 54th WRS, recalls a memorable incident when his crew, in Alaska, was chasing moose down a frozen river bed. "Yes, in a WC-130! I'll never forget the flight engineer on the intercom, yelling over and over, 'watch the wingtips!, watch the wingtips!' . . ."[76] Sometimes we forget that highly trained and disciplined Air Force crews are still boys at heart.

The WC-130 became the "platform" for another automatic computer-based weather observations collection system being developed. This was the Advanced Weather Reconnaissance System, (AWRS), which the National Oceanic and Atmospheric Administration (NOAA) spent millions of dollars to develop and install in a specially modified WC-130B, No. 62-

After Hurricane *Camille*, in 1969, Congress authorized funds to develop an Improved Aircraft Reconnaissance System. Here Lt. Henry Turk, Keesler AFB (1975), works with the Advanced Weather System on board a WC-130. This was the forerunner of the Improved System. *Photograph courtesy of John Pavone*

3492. The idea of the AWRS was to have an automatic navigation system, computer-processed sensor system and high speed radio transmission to a central communications point.[77] Col. Paul McVickar, O'Fallon, Illinois, who served with the 53rd WRS, wrote, "When it [AWRS] worked, it was wonderful. When it didn't, it was a heavy-weight pig with short legs. The system never proved successful because the computer kept failing, and eventually all weather equipment was removed and the aircraft was transferred to the reserves as a "trash hauler."[78] The WC-130C was later sold to Pakistan in the mid-'80s. Air Weather Service's dream of a fast aircraft for hurricane eye penetration at ten thousand feet, with automatic electronic observations, was fitful at best—a tribulation of spent money and crushed hopes.

In the late sixties, with military downsizing, budget problems and the beginning of the satellite era of weather surveillance, the Air Force felt pressed to get rid of expensive weather recon missions. The result was the loss of all WB-50s and WB-47Es and some weather squadrons. Weather missions were costing from $500 to $1000 per hour to fly, and weather recon aircraft became increasingly expensive to maintain, and the Air Force was getting irritable at having to fly non-military missions which were more properly the responsibility of the Air Weather Service or the Department of Commerce, both related to civil rather than military needs. Modification of C-130s to WC-130s came to a halt. Then, in August 1969, a new "lobbyist" hit Washington. A boisterous lady named *Camille* attracted the attention of Congress. Hurricane *Camille* roared across Louisiana, Mississippi, and Virginia, taking 256 lives and leaving $1.42 billion in damage.[79] As a result of *Camille* and public pressure for better hurricane warnings, Congress "found" $8 million to upgrade its hurricane recon fleet. This provided eleven modified WC-130Bs to the AWS and one modified WC-130B to the National Oceanographic and Atmospheric Administration (NOAA) for storm research. This WC-130B, No. 58-0731 was known as "NOAA's Ark."[80]

Severe weather continued to take its toll in lives and property, especially along the population-intensive northeast coastal areas. In February 1969, a howling winter storm dumped sixteen inches of snow on New York City, leaving the city paralyzed. Immediately, NOAA asked the National Weather Service to fly into severe winter storms to gather research data for better forecasting. *Operation Cold Coast* was created, and the *Hurricane Hunters* of the 53rd WRS, in weather modified WC-130 *Hercules* aircraft, began flights to provide observations and timely warnings of ice and snow storms that threaten and cripple the northeastern United States. The first mission, from Ramey AFB, Puerto Rico, was flown on 10 November 1969.[81]

During the first year of *Operation Cold Coast*, the 815th of the 53rd WRS flew 450 hours through the icy storms in fifty missions out of Pope AFB, North Carolina, or Ramey AFB, Puerto Rico. From 1 November 1970 through March 1971, two hundred flights were made, and in the third season, 450 missions were flown—a total of four thousand flying hours, covering 1,334,000 miles in the first three years. The first year of the winter storm missions, 53rd flight crews were from Patrick AFB. After that, relief crews from the 55th WRS, McClellan AFB, California, worked with Patrick crews. One annual report stated that weather observations from the *Cold Coast* crews provided information which changed 16.9 percent of all anticipated forecasts or storm warnings. New York University used the flight data for research to try to discover why winter storms varied so much in intensity and which areas of a storm would yield the most useful information. After several missions, weather recon flight crews considered *Cold Coast* winter storms operationally more dangerous than severe hurricanes and typhoons. The flights encountered constant turbulence and ice, the greatest dread of pilots.[82]

In all of weather reconnaissance, especially in the *Cold Coast* missions, the two most feared flight problems were structural failure from extreme turbulence and loss of flight control because of icing. Turbulence is caused by severe up and down drafts that bounce the plane around like a rough road jolts a speeding car. Pilots fight to control the aircraft, crew members cannot do their jobs, objects are thrown around inside the plane, and everyone hopes the wings will stay on. R. D. Gilmore, Navy flight meteorologist, Oak Harbor, Washington, in severe turbulence, lost a wing tip in a storm. "The wing tip just snapped off during a low level penetration. . . . It felt like someone slammed the bottom of the aircraft with a huge fist." However, the sturdy WC-130 was able to complete the mission and land safely.[83]

Ice, from freezing, moisture-heavy clouds, is one of the most treacherous inflight encounters as it slowly builds up on the wings, coating control surfaces and propellers. Ice adds weight, reduces lift, increases drag, and drastically changes aerodynamic design characteristics of the aircraft in addition to scaring hell out of the crews. The pilot has less control, and, in extreme conditions, loses control. Unless the ice can be removed, or the pilot is able to leave the icing zone, the plane cannot hold altitude—a possibly fatal situation over water or mountains. Some aircraft have inflatable rubber "boots" on the leading edges of the wings and control surfaces. By inflating the boots, if all works well, the ice is cracked and torn off by the wind rushing over the wings. Propellers are protected by a thin film of deicer fluid metered from the propeller hub, and ice does not stick. Often, even these precautions fail to remove certain kinds of ice. On *Cold Coast* missions, it was necessary to fly in the icy soup and risk the dangers of ice buildup to be able to measure the moisture content of the storm. Ice can be deceiving. Weather observer Robert Smith, Huntsville, Alabama, reported that on a routine spiral sounding mission over the Gulf of Alaska, "I was busy with soundings and kept reporting that we were in the clouds since the windshield was totally gray. But something was wrong— our rate of climb took forty-five minutes to go from eleven to twelve-thousand feet. I glanced out the side and saw that we were on top of the clouds and the plane had from four to six inches of clear ice on everything, including the propeller blades—the plane simply wouldn't go above twelve thousand. The aircraft commander tried to get rid of the ice, revving up the props, but nothing worked, until we dropped down to one thousand feet for the ice to melt. We had to abort the mission.[84]

Paul McVickar, O'Fallon, Illinois, was a navigator on *Cold Coast* missions from 1976 to 1980 and encountered both turbulence and ice. Flying in a WC-130E *Hercules,* he describes the missions, "We flew into the center of winter storms—'Nor-Easters,' we called them. I remember being so iced up we were losing altitude, and sometimes we did not think we could reach landfall before we burned off the ice. No one ever crashed from these missions, but icing was really terrible until we began flying 'H' models and were able to measure the moisture with radiosondes dropped from twenty-four thousand feet, instead of flying between five and ten thousand feet. Turbulence was always extremely heavy. . . . At times I could not plot a fix for over an hour, and we were bouncing so badly I could not hold a pencil on one place on the chart."[85]

A similar cold weather project called *Operation Cold Cowl* was flown by the 54th WRS out of Elmendorf AFB, Alaska. These flights provided information on Arctic and Pacific storms moving into Canada and Northwestern U.S. Following the success of fog missions in Alaska in the winter of 1967-68, *Cold Cowl* was continued into 1968-69, and a similar project was initiated in Europe, called *Cold Crystal.* The project was conducted by the 53rd WRS, on temporary duty from Ramey AFB, while the 54th WRS continued research in Alaska. The purpose of *Cold Cowl* and *Cold Crystal* was to attempt to eliminate heavy fog caused by super-cooled water droplets that formed in temperatures from just below freezing down to -20°F. The cold fog was constantly closing air bases and disrupting winter traffic in Alaska and throughout Europe, especially in Germany. Bases in Germany where fog dispersal experiments were carried out included Bitburg, Hahn, Ramstein, Rhein-Main, Spangdahlem, Wiesbaden, and Zwiebrucken. The English base was Mildenhall.[86]

Experience had shown that fog caused by super-cooled water droplets, if supplied with a seeding agent, would freeze and participate as snow. So, similar to cloud seeding, to precipitate suspended moisture, *Cold Cowl* and *Cold Crystal* were fog-seeding missions. Depending on wind conditions, WC-130 aircraft dropped crushed dry ice into the fog. A standard practice for clearing an airfield was for the aircraft to skim just above the fog layer, flying five parallel tracks, three miles long and one half mile apart, seeding at the rate of about ten to fifteen pounds of crushed dry ice per mile. As the dry ice entered the fog, it vaporized and cooled the air to -0°C changing the fog to small snowflakes that fell to the ground. The seeding and clearing process usually took about thirty minutes. In some cases, with wind, the seeding process was accomplished upwind of the airport and, if all calculations were accurate, the wind would carry the cleared air over the runways.

The success of the fog dispersal experiments increased the number of arrivals and departures that otherwise would have been canceled. In the first three years, *Operation Cold Cowl* allowed an average of 153 arrivals and 170 departures that would have been delayed or diverted; *Cold Crystal* prevented delays or cancellations of an average of 54 arrivals and 87 departures.[87] Whether the results justified the expense of specially equipped aircraft and trained crews was a matter of concern although it was recommended that the project be continued.[88] *Cold Cowl*, which started in 1967, was continued into 1971 and 1972 by the 11th WRS at Elmendorf AFB, Alaska.

Dry ice seeding was expensive in aircraft, time and crews, and another system of fog control was sought. Since 1966, Orly Airport, near Paris, France, had been using a system of fog dispersion using propane gas, and some airports in the U.S. and Germany were also using this system. Propane from portable spray units operating at ground level was released at various points around airports. The system was much less expensive— $13,500 per year for the propane as opposed to $360,000 per year for airborne seeding. Preparation and response time was much faster. Aircraft required one hour to respond as opposed to five minutes for the propane system which could be activated from the ground by remote control. The propane system was adopted and is currently in use today at fog-prone airbases in Alaska, Washington, and some European fields.[89]

7

Satellites and the Era of Drawdowns

As we moved into the seventies, weather recon planes were flying missions in all parts of the world. Somewhere, day and night, a weather plane was in the air. But inexorable forces such as budgets, technical progress, warming of the cold war, fretting of the sheiks of OPEC, and the prejudice and prudence of the military were shaping the future of air weather reconnaissance. To put the events in this chapter in perspective, we must go back in history to seemingly unrelated happenings. Shortly after World War II, on 13 October 1959, a tiny "cloud" appeared on the weather horizon—just a wisp of a cloud—probably nothing that recon should be concerned with. On that day, the United States placed its first dedicated meteorological satellite, Explorer II, into orbit. This research satellite, with instrumentation, including an infra-red radiometer, was used in the detection of meteorological phenomena and relaying weather imagery to earth.[1]

On 1 April 1960, a Television Infrared Observation Satellite (TIROS-I) was launched. This more sophisticated satellite gathered information on a global basis for weather forecasting. TIROS-I's instruments could also track aircraft and ships at sea and the movement of marine life and pollution.[2] TIROS-I made the future of weather reconnaissance darker and more threatening—the forecast of a gathering storm like no other storm the recon crews had conquered before—that would cut the switches and cancel the flight plans of the mighty world-wide weather reconnaissance fleet.

As the cold war warmed, and with Congressional emphasis on reducing the military budget, six recon squadrons—the 2078th, 313th, 414th, 373rd, 374th and 375th—had been deactivated in 1950 and 1951. On 8 May 1964, the 59th WRS was inactivated, and on 30 November 1969, the 57th WRS stood down. The weather fleet was losing its numbers, and the decade of the 1970s became known as the "Era of the Drawdowns," a constant battle between the Congressional budget cutters and the Air Weather Service. Those attempting to cut the weather services were called the "Dirty Dozen," and John Fuller, weather historian, called the attempt to stop the "Dirty Dozen" akin to "rearranging the deck chairs on the Titanic."[3]

Historians have recognized the "global village concept," in that what happens in one part of the world profoundly affects what happens in another seemingly far removed part of the world. In the mid 1970s, two events contributed to the death of weather reconnaissance as we knew it. The first was the Arab-OPEC oil embargo, in 1973, which shot POL (Pentagonese for petroleum, oil and lubricants) costs into the stratosphere. This inflation caused the Air Force to take a hard look at all missions, eliminating or drastically reducing those which were peripheral to basic wartime needs. William V. Yelton, Cañon City, Colorado, former action officer for weather matters on the staff of Hq. Air Weather Service/Air Operations (AWS/AO), describes a "wall-to-wall" special study, in 1974, at Hq. Military Airlift Command (MAC): ". . . the study . . . forced each staff element to fall back to zero base and add only missions which had a basic charter at Air Force or Department of Defense level, eliminating those tasks which had been added by lower authority."[4] Thus, if a mission had no basic charter authority, even if it had been flown for many years, it had no authority to be flown. This new order forced many aircraft into mothballs and eliminated flights of "boring holes in the sky" just to get flight time.[5] As the war in Vietnam came to a close, MAC's Air Rescue Service (ARS) suddenly found itself with a group of highly skilled aircrews and support personnel, many aircraft, and only a skeleton mission. The Air Weather Services recon resources were transferred to ARS to keep that unit's mission alive.[6]

As the weather forces dwindled, and weather satellites assumed more responsibility for global reconnaissance, atomic air sampling continued to be the primary mission. The Air Force continued its search for the ideal aircraft. The Boeing WC-135, a converted tanker, was selected for atmospheric sampling because of its size, speed, and cost-effectiveness. The aircraft was occasionally used for storm and cyclone reconnaissance although, as with the other jets, the WC-135s were not recommended for typhoon or hurricane penetration and were never used in or near combat areas. Ten WC-135Bs were transferred to the Weather Service, beginning on 2 August 1965 until 21 January 1966, five each to the 55th and 56th WRSs.[7] The WC-135 was used only for "special assignments."

Missions of the WC-135s were world-wide and varied. For example, on 25 June 1967, Maj. Jerry Fuller of the 55th WRS and his WC-135 weather crew were completing a forty-five day weather project in Argentina. On their return flight to McClellan, they were asked to divert their flight to fly a storm sortie into Hurricane *Carlotta* in the Atlantic. "After all," the request said, "the storm is only an hour off your regular course." Heading home after seven weeks, even an hour's delay can be almost unbearable. The WC-135B had never flown a tropical storm, and, although the pilots had had previous storm recon experience, they wanted this special distinction for their aircraft. They aimed for the point where the satellite said the storm would be, found the eye, both visually and by radar, flew over the hurricane, rechecked its position, movement, wind direction, and speed, and released three dropsondes into the storm center. After taking photographs and final observations, they radioed, "We came, we saw, we computed," and continued their flight to McClellan as the first WB-135 to "fly" a hurricane.[8]

The remaining weather squadrons, both Army and Navy, were still flying and had begun providing a new and unique service to the space program of weather scouting for pre-launch, down-range and splashdown areas. On 23 July 1969, the day before splashdown of Apollo 11, a weather satellite and Navy aircraft reconnaissance had indicated heavy showers and thunderstorms in the recovery area. After coordination with Navy's Weather Central and the Weather Bureau, the National Aeronautic and Space Administration (NASA) recommended the landing area be moved some two hundred miles, and Neil Armstrong, Buzz Aldrin, and Apollo 11 came down in good weather.[9] Support of space missions was an important diversion from atmospheric sampling and storm missions. Crews of the 53rd WRS at Ramey AFB, in Puerto Rico, flying WC-130s, covered Cape Kennedy blastoffs and the 55th WRS WC-135 crews, from McClellan AFB, California, patrolled landing areas in the Pacific.

On 19 April 1968, Ralph Plaisted, an insurance executive from Minnesota, and three other amateur explorers reached the North Pole by snowmobile—the first to make the journey by motor. On that day, Leslie Nelson, 55th WRS weather technician, Sierra Vista, Arizona, and his crew, on a return flight from Eielson to the North Pole, were asked to verify that the expedition had reached the pole. Flying their regular mission, in a WC-135B, the pilot dropped down to eight hundred feet, located the four men and, after navigators had confirmed their position at the pole, radioed to expedition

The Air Force selected the Boeing WC-135, a converted tanker, for cost and dependability. More friendly than other jet planes the WC-135 was stressed for storms although it seldom was used for hurricane penetration. It served weather for eighteen years. *Photograph courtesy of Clarence Miller*

command headquarters: "Party was at the North Pole."[10] *National Geographic Magazine* called these explorers, "the first to have their success verified; a U.S. Air Force weather plane recorded their presence."[11]

The WC-135 was a more "friendly" aircraft than previous jets. On 7 July 1969, Maj. Henry M. Dyches, pilot, was on takeoff roll from Yokota, Japan, when a problem caused the flight controls to bind, and the plane could not be rotated to take off attitude. In a split-second decision, using the elevator trim control, the pilot made a successful takeoff. Then by using trim, differential spoilers and throttle, Major Dyches was able to make a successful "go around" and an emergency landing in minimum weather. Many lives were saved, and the pilot received the Korean Kolligan, Jr. Trophy for 1969.[12] In eight and one-half years, Boeing built 820 of the four-jet tanker aircraft, including eighty-eight variations, the WC-135 being one. Even at a cost of some $2 million each, they were the lowest cost-per-pound of airframe of any military aircraft in production—820 planes at a cost of $1.66 billion—an exceptional value for the Air Force.[13]

Explorer Ralph Plaisted, in April 1968, set up camp at the North Pole. This first journey to the pole by motor was verified to the National Geographic Society by a 55th Weather Recon Squadron WC-135 aircraft. *Photograph courtesy of Leslie Nelson*

The WC-135 was tougher than other jets. Leslie Nelson was on a "storm stress" mission to test the WC-135 in turbulence. The aircraft was wired with stress sensors and flown in storms on the eastern slopes of the Rocky Mountains. The test required "flying in the most extreme turbulence for seven to eight hours—that makes for a real fun flight." The aircraft stood all the stress. Nelson also reported a twenty-five hour mission, scouting radioactive debris from a French nuclear test. They flew round-robin from McClellan, California, around Tahiti and got into a very "hot" cloud. The crew had to go on oxygen with no smoking, food, or drink for the rest of the mission. The pilot flew through all the rain clouds he could find to wash the aircraft and on landing it was still hot. After several ground washings, the 135 had to sit on the ramp for over a week before it lost enough radioactivity to be safe for another mission.[14] From July 1972 until August 1993, the WC-135s were phased out, some sent to mothballs and other modified for other duty in the Air Force, thus ending eighteen years of honorable service to weather.[15]

Ralph Morgan, Homer, Alaska, dropsonde operator of the 55th WRS, describes his experience with reconnaissance for Apollo 13, the mission that was aborted and came back for an early splashdown in the Pacific: ". . . our WC-135 crew, on a regular mission from McClellan to Eielson, Alaska, was diverted to Hickam, Hawaii, and put on 'crew rest' the moment we landed. Four hours later, we were alerted and flew to Pago Pago, American Samoa, along with another WC-135 from the 56th WRS, Yokota, Japan. We flew a weather track enroute. . . . Our B-4 bag had been packed with Arctic gear for the Alaskan mission, and the heavyweight clothes didn't fit the climate of Pago Pago. We did get to see the astronauts when they were flown in from the recovery ship."[16] The importance of the weather missions and good forecasts was impressed on NASA in 1970 when lightning struck Apollo 13 during liftoff.[17] Maj. Dennis Werking, part of the weather crew supporting Apollo 14 and 15 missions, wrote, "It was the best seat in the house; situated forty miles away, at twenty-two thousand feet, we got a fantastic bird's eye view of the blastoff, the first-stage separation, and the second-stage ignition."[18]

Although enthusiasm for space glamour missions was high, and lift-off and splashdown assignments were eagerly sought by weather squadrons, the more perceptive weather crewmen may have realized that the research and development of satellite systems would eventually replace aircraft observations,—electronic "eyes in the sky" would take over weather surveillance. Actually, an early prediction had been made in late 1959: in "NASA Weather Prophecy," *Air Service Observer*, December 1959, prophesied ". . . in the next decade . . . it seems possible that a fully operational meteorological satellite system having world-wide coverage will be in operation. . . . 'stabilized satellites' in polar orbits five hundred to one thousand miles altitude, and 'stationary satellites' in equatorial orbit at a 22,000-mile altitude . . . feeding information into a national weather center."[19]

Budget problems, and the shift of national priorities in the search for Vietnam "peace dividends," were as effective as the new weather satellites in grounding the weather program. President Nixon's appointment of Gen. Jack Ryan, as Air Force Chief of Staff, in early 1969, may have been the fatal blow. General Ryan was not a friend of weather reconnaissance and believed, as many AWS officials secretly knew, that weather reconnaissance's emphasis on synoptic missions did not increase the accuracy of global forecasts, and weather reconnaissance was expendable.[20]

Col. Paul E. McVickar, chief of programs division of the U.S. Transportation Command, Scott Air Force Base, Ill., describes the forces that brought weather from a mighty fleet to its current one reserve squadron at Keesler AFB, Mississippi—a parade of relentless changes. Colonel

McVickar outlines his history, not as a "Monday-morning quarterback," but as a nostalgic historian who served his time in the air as a navigator with twenty-five hundred hours in weather reconnaissance, fifty-five hurricane and typhoon penetrations, and as project director for Hurricane *Gilbert*, the strongest hurricane ever flown in the Atlantic.[21]

Why was the Air Force trying so hard to eliminate the hurricane hunters? Prior to 1975, the AWS, at Scott AFB, Illinois, owned the aircraft, crews, squadrons, etc. The subordinate unit was the 9th Weather Reconnaissance Wing, McClellan AFB, California. By 1975, satellite technology had replaced the need for daily synoptic flights to 'locate' suspect areas, and the number of squadrons had been reduced to three—the 53rd at Ramey (Puerto Rico), 54th at Anderson (Guam), and the 55th at McClellan (California). The 53rd and 54th flew exclusively WC-130s, while the 55th had both WC-130s and WC-135s.

The AWS was concentrating its dwindling funds on satellite technology and traditional base weather stations. Aircraft and recon crews were becoming an unnecessary management headache and fuel costs were straining every budget. Reorganization was in order, so the 53rd moved to Keesler AFB, Mississippi, when Ramey was closed. The 54th stayed at Anderson, but rather than a colonel for commander of the squadron and the maintenance unit, the new commander was a lieutenant colonel.[22] This lower rank was a small distinction, but, to the military, it was a smoke signal that indicated a curse had been put on weather recon.

The 55th transferred all of its WC-130s to the 53rd WRS and to a newly created-reserve unit, the 815th WRS, part of the 920th Weather Reconnaissance Group. The 9th Weather Reconnaissance Wing cased its flag forever, and a new wing, the 41st Rescue and Weather Reconnaissance Wing, was created on 1 September 1975. At this point the weather recon fleet consisted of twenty-one WC-130s and seven WC-135s—seven WC-130s to each of the 53rd, 54th and 815th WRSs, and seven WC-135s to the 55th.

Typhoon reconnaissance was continuing and small personal stories add to the temper and character of weather flights—seemingly insignificant dramas that never reach official history. The story of WC-130H, No. 65-0965, tells of the only WC-130 ever lost in weather reconnaissance. Col. Paul McVickar records this experience: "It happened on 12 October 1974, one month after I arrived in the 54th WRS, Guam. . . . My family and I had settled on a house in Daededo, Guam, when my wife heard some squadron wives talking about a lady being raped in that subdivision. She told me to find a better location. I went to the scheduler and asked to be taken off a flight leaving the next day to fly Typhoon *Bess*. Fortunately, I was not qualified in the aircraft and was taking a training flight, so I simply was taken off the orders, and the crew departed with my instructor in the navigator position. The aircraft disappeared during the night somewhere in the South China Sea between fixes—no cause of the accident was ever found, nor was any part of the wreckage. The official accident report stated, '. . . the aircraft took off and never returned.'"[23]

In 1975, the military was not immune to the fears of aircraft highjacking, and members of air crews often carried sidearms. This incident happened on 10 June 1975, when crew chief Robert Styger, Willingboro, New Jersey, and his crew were returning to McClellan AFB, California, in a WC-135B, from a sampling mission near Japan. "We were on our way back from Japan and dropped in on Midway to pick up a few Navy and Air Force folks who needed a ride back to the states. In those days two crew members carried a gun because the threat of 'highjacking' was high. We checked all baggage of non crew and each new passenger. While checking one of the Navy people, we found bullets in his bag and, after searching, he had a gun! It was some sight to see our gun-carrying crewmen grab this guy and pull him away from the aircraft. We then found he was a part of the base shore patrol. However, before he was allowed to board, his gun and shells were put up front with the navigator."[24]

Navy weather recon was also standing down. On 1 July 1971, Navy Airborne Early Warning Squadron One was disestablished, and on 30 April 1975, the famed Navy *"Hurricane Hunters,"* the squadron that had shared missions with the Air Forces' 53rd WRS for thirty-two years, was discontinued. The following official directive was issued:

R 301304Z APR 75
FM WEARECONRON FOUR - TO CNO WASHINGTON DC—
    INDO CINCLANT FLT NORFOLK VA—
    COMNAVWEASERVCOM WASHINGTON DC—
    COMNAVWEASERFAC JACKSONVILLE FL—
    COMSEABASEDASWWINGSLANT JACKSONVILLE FL—
    NAS JACKSONVILLE FL—CHINFO, WASHINGTON DC
BT WEARECON FOUR DISESTABLISHED—A. CNO 122137Z
    FEB 75
1. THE WORLD FAMOUS HURRICANE HUNTERS, THEY FLY
    NO MORE
WE HAVE DISESTABLISHED WEATHER RECONNAISSANCE
    SQUADRON FOUR
WE'VE RANGED THE GLOBE SINCE FORTY THREE
IN SEARCH OF STORMS THAT CROSS THE SEA.
TWO THOUSAND TIMES WE'VE CROSSED THE EYE,
IN DARK AND WET AND STORMY SKY.
TWO THOUSAND TIMES WE BRAVED THE GALES

FOR MOTOR SHIPS AND THOSE WITH SAILS.
THREE HUNDRED STORMS OF FEARFUL WRATH
TO WARN THE PEOPLE IN THEIR PATH.
FOR COUNTLESS HOURS THESE FEARLESS MEN
    FLEW PLACES WHERE FEW MEN HAVE BEEN.
WE LEAVE A RECORD FOR THE FLEET TO CHASE
A SEVENTEEN YEAR ACCIDENT-FREE PACE.
AND NOW IN ACCORDANCE WITH REFERENCE (A),
THE HURRICANE HUNTERS ARE NO MORE TODAY.
BT[25]

   In bitter verse the Navy closed its hurricane reconnaissance mission, a saga that began in 1943. While the Navy reconnaissance crews were active for thirty-two years, their missions were more specific to Navy needs, especially their ships at sea. One Navy weather history lists the first reconnaissance and hurricanes flights beginning ". . . early in 1943, '44 and '45 [when] the Army and Navy carried on a project of research and observation of hurricanes in the Caribbean and Gulf of Mexico. In 1943, in Miami, the Joint Hurricane Weather Service Office was established . . ." However, most of the early history regarding these flights consisted of vague references in other unit's histories, and there is no recorded information. Some 'hunters' from the early days recall that Martin Patrol Bombers (PBMs) were involved, presumably from Naval Air Stations where PBMs were assigned. . . ."[26]
   In the thirty-two year summary of Navy recon flights, three hundred storms, including 201 hurricanes and typhoons were flown although no penetrations were made until 1948.[27] During the Navy years of storm chasing, there were twenty-five "Summer Hurricane Detachments" in Florida and Puerto Rico. Navy "wind chasers"[28] had long productive years of service and the definitive history of their weather reconnaissance flights has not yet been written. A personal note: I remember, in those first early days of weather recon, in 1944, flying across the North and South Atlantic, with our lack of experience in Arctic weather and the awesome job of trying to find and track Atlantic hurricanes, we "thanked God there was a Navy," and were relieved to know that "out there somewhere" was a Navy crew fighting the same icy clouds, the same winds with the same anxiety. The Navy, like the Army "flew the missions that no one else wanted to fly and flew the weather which nobody would fly into."[29]

   As part of the reorganization, the Air Weather Service (AWS) transferred aircraft, crews and squadrons to Aerospace Rescue and Recovery Service (ARRS)—this made good management sense, since the WC-130H models were converted HC-130H aircraft. However, AWS did not transfer funding. At first this was no problem since AWS and ARRS were funded

As the 920th Weather Recon Group (AFR) WC-130 completed its first hurricane mission, on 9 June 1976 into Hurricane *Annette*, the crew was greeted by the group commander, a congressman, the plane's ground crew, and the base commander. *Photograph courtesy of Ralph Morgan*

as Major Force Programs (MFPs) in the Department of Defense (DOD) and paid for from what was called Base Operating Support (BOS). Billing from BOS was the first paid—almost automatic, like turning on the light switch. However, by 1986, major budget reduction was forcing the Air Force to start cutting structure. The AWS was interested in using the $25 million needed to maintain the 53rd and 54th WRSs to make up for budget cuts. AWS wanted to keep its satellites in orbit and to maintain base weather stations. Aircraft reconnaissance became a drain.[30]

On 1 September 1975, the Air Weather Service weather recon mission was transferred to the Aerospace Rescue and Recovery Service (ARRS),[31] and the 53rd, 54th, and 55th WRS were assigned to ARRS, a sort of "holding company" for unwanted weather squadrons. For the first time ever in history, no member of the Air Weather Service command (chief of staff, commander, or vice-commander) held an aeronautical rating[32]—the death knell for flying crews.

As heir to hurricane reconnaissance, the Air Force Reserve 920th Weather Reconnaissance Group (WRG)[33] flew its first hurricane mission from Keesler, AFB, Biloxi, Mississippi, on 9 June 1976. The reserve crew, heralded as "local Gulf Coast citizens" by a Biloxi newspaper, flew a WC-130 Hercules into Hurricane *Annette* off the southwestern coast of Mexico. The crew made two penetrations into the storm as it threatened the touristy coast of Mexico before it moved off into the Pacific. The 920th played a major role in storm detection in the 1976 season. Crew members on this first reserve flight were Lt. Col. Walter Craig, aircraft commander; Majs. Fred Floss, weather officer, and Jack Hansen, navigator; Capts. R. E. Shepherd, pilot, Kenneth Gates, pilot, and Gilford Tures, navigator; 1st Lt. William Sims, pilot; M/Sgt. W. P. Meins, flight engineer; and T/Sgt. Ralph Morgan, weather observer.[34]

Even as weather reconnaissance was being put to rest, personnel changes were being made. As late as the early seventies, "men only" signs were figuratively in place at all weather aircraft flight lines. Women had not been accepted for positions in the air. This exclusion seemed unusual, since, in November 1943, fifteen Women Air Force Service Pilots (WASPs), who flew millions of miles throughout the world, were assigned to the Air Weather Service as administrative and ferrying pilots. WASPs were women civilian pilots, directed by Jacqueline Cochran, famed woman pilot. However, the program was discontinued in December 1944, and the only women in the weather service were Women's Army Auxiliary Corps (WAACs), serving as meteorologists and weather ground staff. The military was not open to women in combat roles. Many weather commanders, with honest sincerity, believed that aircraft duty was too strenuous for women, that long missions with men and women working together in cramped quarters, and the absence of "his" and "her" personal facilities would cause problems that might reduce the effectiveness of reconnaissance missions.

However, the winds of sexual equality continued to blow, and the first woman flight crew member, in 1973, was Sgt. Vickiann Esposito, dropsonde operator for the 53rd WRS, at Keesler AFB, Mississippi. Sergeant Esposito later served with the 54th WRS at Guam and earned the Air Medal for fifteen typhoon penetrations.[35] As the first female application for airborne weather duty was being considered, the decision for some was painful. Many military commanders were reluctant to change things that were working and underestimated the pressure of social change. However, early in 1972, Hq. AWS-Air Operations learned that a "female NCO,"[36] Sergeant Esposito, had applied for duty in the reconnaissance program, and a personnel meeting was called to consider the application. While most personnel officers were willing to approve the application if Sergeant Esposito met all qualifications, there was some opposition. Sergeant Esposito was

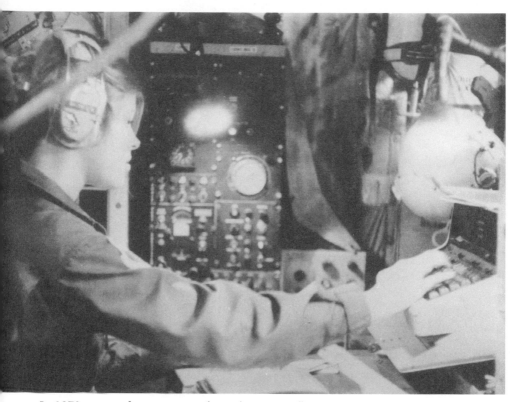

In 1973 women became part of weather recon, flying in every crew position. Here Ann Heather Ford, dropsonde operator, works at her position. One commander could have flown up an all-female crew although the women vetoed the idea because of the excessive publicity it would have brought. *Photograph courtesy of Paul McVickar*

qualified and had passed a regimen of physical tasks. Before the meeting, a colonel in MAC headquarters called personnel to say that "MAC headquarters is dead set against opening weather reconnaissance to female applicants."[37] Should she be accepted, it would set a precedent. It was generally expected that the board would disapprove the application, and the meeting would be "just going through the motions." This attitude on the part of the MAC staff was easy to understand since Air Force policy at that time prohibited women in combat roles, and weather reconnaissance had a world-wide mission with a "combat ready" status. Missions were often near or over enemy territory, and some weather flights had been intercepted by Soviet aircraft, setting up a quasi-combat situation.

Col. William V. Yelton, on the staff of Hq. Air Weather Service/Air Operations, stopped the discussion by reading a policy letter by the Commander of MAC on "Equal Opportunity and Treatment" (EOT). "No one on the MAC staff would go on record saying that COMAC didn't mean exactly what he said and Sgt. Vickiann Esposito's application was approved." Colonel Yelton said.[38]

Sgt. Carolyn Rowe became the first female WC-130 flight engineer with the 53rd WRS. First Lt. Nancy Holtgard was the first woman to qualify as an aerial reconnaissance weather officer. She flew with the 54th WRS at Anderson, Guam, from 1976 to 1978. First Lt. Flo Fowler was the first woman navigator and was assigned to the 53rd from December 1977 until she moved to the 54th WRS in March 1981 for a three-year tour. The first woman pilot qualified for reconnaissance was Lt. Carol Scherer, member of the first Air Force class of eighteen female pilots. Lieutenant Scherer, after graduation, was assigned to the 54th WRS to fly WC-130s. Capt. Valerie Schmid and Capt. Nola Cary, as of 1995, were still flying with the Air Force Reserve recon squadrons.[39] The "men only" roles had been changed.

John W. Pavone, Chief of Aerial Reconnaissance Coordination, All Hurricanes (CARCAH), in Coral Gables, Florida, wrote, "While in charge of the AWS detachment of the 54th WRS, . . . I had five female crew members. . . . They were all a cut above, and knowing they were a minority, they tried harder. . . . I am proud of what they accomplished after their reconnaissance assignments. Carol Belt and Candis Weatherford received Air Force-sponsored PhDs, and Belt was selected for astronaut training. Marsha Korose and Beverly Baker received graduate degrees in meteorology. Nola Cary, dropsonde operator, earned a degree in meteorology and was commissioned an officer in the Air Force reserve. She is flying hurricanes today [1995].

"Since Sergeant Esposito broke the male barrier in 1973, all weather reconnaissance squadrons have had women crew members and they have filled every crew position. In fact, Col. Paul McVickar, as director of operations for the 53rd WRS in the late 80s, could have launched an all-female crew. The women themselves decided that they did not want to fly as a unit because of the publicity the flight would have generated. In their own words, 'we prefer to be crew dogs.'"[40]

Maj. Gen. John Collens, USAF retired, president of the Air Weather Association, and former commander of an air weather squadron, wrote, "Eventually, only the Atlantic east coast and Gulf coast and Caribbean area retained a weather reconnaissance mission, albeit under the ARRS." The political clout of East and Gulf coast members of Congress and the National Hurricane Warning Center kept the mission in that area and financed it through the DOD budget."[41]

The 53rd had been cut from the Air Force budget in 1986, '87, and '88, but each year, the Congress reinstated it. The 54th was not so lucky. It was cut and closed in 1987. The 54th was fairly easy to cut because it was a strictly military unit with one military customer—the joint Typhoon Warning Center operated by the Navy Oceanographic Command.[42] The "customer" concept was a new idea for financing weather recon—sell recon services to a customer who wanted the information. "There was a common thread throughout, namely, find out what the customer wants, figure out how to satisfy his needs, and then handle the data electronics, etc., promptly."[43] But this did not generate enough funds to keep the aircraft flying.

Drs. Neil Frank and Bob Sheets, of the National Hurricane Center, masterfully kept the 53rd open and the recon aircraft flying by appealing to the public and Congress. Every time the Air Force announced that the 53rd had been cut, the National Hurricane Center would go public with devastating predictions of storm damage that could happen if they did not have weather reconnaissance data. The only way the 53rd was finally closed in 1990 was when Congress directed the Air Force to transfer all WC-130s to the Air Force Reserve and maintain weather recon on a reserve basis. All twenty WC-130s were transferred to the 815th. The reserves reactivated the 53rd WRS, still nicknamed the 'Hurricane Hunters,' and they continue to fly hurricanes as a reserve unit.[44] The threat of massive storm destruction in heavily populated areas of the U.S. and the need for precise measurements of the hurricane force have kept the Keesler AFB, Biloxi, Mississippi, in business to date (1996). Previously discontinued, the 53rd WRS was reactivated at Keesler AFB on 1 November 1993 and a public ceremony held in January 1994. In essence the Department of Defense is paying for a mission that is the responsibility of the Commerce Department, i.e. weather warnings to the U.S. public.

The Soviets continued to provoke the U.S. in the unfriendly rivalry of the cold war. On 10 November 1989, Pentagon spokesman Pete Williams said that U.S. officials suspected the Soviets of shooting lasers at American planes in the Pacific, and in one of four incidents, damaging the eyesight of an Air Force crewman. The report stated that two Soviet vessels "may have" aimed lasers at U.S. aircraft on 17 and 28 October and two occasions on 1 November 1989 over waters off Hawaii. Shipborne lasers are designed to gauge distances to guide weapons, but military experts say they can be used to blind or disable pilots of aircraft. The laser shots were made in an area where Soviet ships observe the splashdowns of their missile tests, and American aircraft usually were present to watch movements of the Soviet ships. An agreement to prevent such military encounters was signed in the summer of 1989, but did not go into effect until 1

January 1990. However, such attacks apparently had been expected because Air Force crews were provided with laser eye protection although it was not explained why damage occurred with the equipment in use.[45]

In what might have been another incident of Soviet "friendly" aggressiveness or a downright hostile act, the Air Force announced in a "Weather Reconnaissance Update" letter of 14 September 1989 that a Soviet AN-12 aircraft flew two hurricane tracking missions through Hurricane *Gabrielle* "last week" (7 September). The missions were flown from Cuba, and on the first mission the Soviet presence in the storm was only accidentally discovered. The Chief of Aerial Reconnaissance Coordination for Atlantic Hurricanes (CARCAH) overheard radio traffic between the AN-12 and San Juan, Puerto Rico, air traffic control. At the same time, aircraft from the 53rd WRS were also flying in the same storm. Flying a hurricane is risky enough, but to have another aircraft at an unknown location and altitude is a worse threat than the storm.

The Soviets had also flown hurricane "research" missions from Cuba in 1988. During the winter of that year, Dr. Robert Sheets of the National Hurricane Center (NHC) made arrangements with the Havana weather bureau to notify NHC when the Soviets were going to fly weather reconnaissance missions so that there would be no airspace conflicts. However, on the first 1989 mission, the Cubans did not get through to the hurricane center, but they did establish contact on the second. While the Soviets were within their rights to fly into the hurricane from Cuba, international air traffic control, for safety, demands that all flights in the same area be coordinated. Having heard the radio traffic, CARCAH was able to coordinate altitudes directly with the Soviet recon aircraft and prevent conflict. The San Juan Center later agreed to call the National Hurricane Center when the Soviets were airborne.[46]

As "D-day"—final deactivation—for all weather flights relentlessly moved closer, many people became concerned that the nation would be left with inadequate warnings for hurricanes and the severe northeast winter storms. With millions of the nation's population living near the coasts and the multi-billion dollar buildup of homes and industries within a storm's path, weather warnings were vital. Although the storms could not be stopped or deterred with warnings, preparations could be made and, as the storm was bearing down, there was comfort to hear a radio report that "hurricane recon aircraft, this morning, fixed the storm at. . . . " As weather satellites decreased the need for airborne reconnaissance, the debate involving weather experts, the military and a budget-conscious Congress continued. Was there a need for both satellite and manned flight through the storms? The answer from the NHC was always "yes!"

Yet, the Air Force continued its determined program to drop all aircraft reconnaissance, and the two WC-130 units were threatened with deactivation. Congressional representative Tom Lewis, Florida, opposed the deactivation, saying "This is ill-advised and, if carried out, would literally leave Florida [and all coastal areas] naked and defenseless in the path of a fast-moving killer hurricane."[47]

General Collens, veteran reconnaissance commander, added perspective to this argument, "Except for 'feeling the pulse' of a hurricane/typhoon by dropsonde and eyewall penetration, the location of the storm by satellite can be done very well. Before satellites, weather recon crews spent hours searching for the storm. Now they fly straight to its known location, penetrate and measure the maximum winds. Satellite photos have catalogued various intensities of storms, so the 'approximate strength,' based on photos and actual aircraft reports, correlate very well. As advanced satellite sounders improve, the future need for weather recon cannot be known. . . . But, as we learned from *Andrew* (1992), when major population centers are in the path of the storm, aircraft penetration still gives the best estimates of the maximum winds to be expected."[48]

While debate continued on the future of hurricane hunters, the 53rd moved ahead with the installation of the new equipment for reconnaissance. A new Self Contained Navigation System (SCNS) and the Improved Weather Reconnaissance System (IWRS) were installed in their aircraft, so that no matter what organization might end up flying the WC-130s, these new systems would contribute to more accurate forecasts and warnings that might save lives.[49]

No matter how much influence the great powers of Congress have in budget cutting, a storm that threatens the nation's coasts has greater influence, and in September 1989, Hurricane *Gabrielle* was a powerful "lobbyist." *Gabrielle* may have helped win House approval to keep the recon airplanes flying for five more years. The huge storm brushed near the coast, and alerts were issued from central Florida to New England. As *Gabrielle* approached, aircraft hurricane penetrations showed that the storm's pressure was lower and the winds stronger than estimates from the weather satellite. This provided information that Dr. Robert Sheets, director of the National Hurricane Center, needed to reinforce his argument for aircraft recon: "We require that information [pressure and winds] for warning and forecast purposes. If we don't have [it], we have to over warn."[50]

Hurricane *Hugo* reinforced *Gabrielle's* pressure on Congress. *Hugo*, September 1989, the strongest storm to hit the U. S. since *Camille* in 1969, moved across the Caribbean islands and stormed ashore into South Carolina, taking twenty-eight lives in the Caribbean and twenty-one lives in continental U.S.—remarkably low considering the widespread destruction.

Property damage totaled $10 billion—$7 billion in the U.S. and $3 billion in the northeastern Caribbean.[51]

Dr. Sheets, in reasoned public appeals, continued to emphasize the value of aircraft reconnaissance at Congressional hearings, through the press, and through exhaustive study and research. Dr. Sheets was concerned with "over warning," i.e. predicting a wider area of entry than necessary. There must be great sensitivity in the warning and response process, and Dr. Sheets explained his problem, assuming a storm (such as Hurricane *Andrew*) is moving toward the Florida Keys: "Now the Kennedy Space center is calling because they have the shuttle on the launch pad—they have five of their rockets on their launch pads. What am I going to do? It takes forty-eight hours . . . to prepare for this storm." And they said, "We've got all this out there." And I said, "Yes, and we have four million people in South Florida, too, that are all wanting the same questions answered."[52]

Dr. Sheets continued, "Putting up a warning is a $50 million decision. That's what it costs to prepare the average coastal zone for a storm. Whether the storm strikes or not, once we put out the warning, the public, individuals, the government, etc., spend an average of $50 million preparing for that storm. Now if [the warning] includes a major metropolitan area, like Miami, the numbers go up . . . if it includes, in Texas, the Houston-Galveston area with the petrochemical industry and the large population there, it's probably $80 to $100 million just to prepare those communities. . . ."[53]

Evacuation is a major disruption. Except for those who have experienced a major storm, people are reluctant to leave their homes, and those in "barrier islands," such as Padre Island, Texas, are reluctant to evacuate under "blue-sky conditions." People don't like to leave their homes until they can see the actual threat of rising waters and increasing winds, and, by then, highways may not be capable of handling the crush of traffic, resulting in people trapped as waters cut off escape routes. This situation nearly happened for about two hundred people on western Galveston Island during Hurricane *Alicia* in 1983.[54] Although preparations are expensive, inaccurate warnings can be as much as a thousand times more expensive. In a theoretical model, a hurricane that might strike Miami and Fort Lauderdale squarely could cost insurers $100 billion—and a wind speed increase of only five knots could double the losses.[55]

The Air Force continued to try to downscale storm warning reconnaissance. In May 1992, at a budget hearing, Air Force Secretary Donald B. Rice announced the 1993 budget included fewer planes than Congress requested to track storms in the Atlantic, Caribbean, Gulf of Mexico and the Pacific. Rep. Bob Livingston, from Louisiana, reminded Secretary Rice that the lawmakers were extremely interested in the Air Force Reserves (the 815th WRS of the 403rd Tactical Airlift Wing, at that time, at Keesler AFB, Mississippi). So intense was the interest, lawmakers specified that the reserve unit should have ten aircraft assigned to the mission and two back-up aircraft, with fourteen full-time crews and three part-time crews available to fly sixteen hundred hours.[56]

Secretary Rice said that the Air Force had hoped that the Coast Guard or National Oceanic and Atmospheric Administration (NOAA) would pick up the missions, but admitted that the service was no longer trying to shift that responsibility. Rep. Livingston said, "The Air Force does a good job, and its pilots are superbly qualified, [but] the overhead technology is inadequate to maintain constant tracks . . . when we're talking about the possibility of ten- to fifty-thousand lives being jeopardized. . . . [I] think the total public interest would be better served by the Air Force keeping this mission."[57]

The United States experiences more hazardous weather conditions than any other country in the world, averaging ten thousand violent thunderstorms, five thousand floods, one thousand tornadoes and several hurricanes each year.[58] Hurricane *Andrew* (1992), the third strongest storm on record, caused more damage than all of the earthquakes this century (to that date) in the continental United States. Yet in Congressional action, lobbying and political self interest, things often get skewed. It was reported at the 1993 National Hurricane Conference, the Federal Emergency Management Agency (FEMA) spends $50 for earthquake related programs to $1 for hurricane related programs, excluding relief efforts.[59]

At the end of the 1995 hurricane season, the nation was reeling from the worst storm season since 1933. "No one under retirement age can recall a hurricane season quite like this," Dr. Bob Burpee, new director of the National Hurricane Center, said. "Twelve tropical storms have materialized off the coast of Africa and six have grown into full-fledged hurricane. The good news is that the damage . . . has been comparatively mild. The bad news is that more big storms are on their way. . . ."[60] Atmospheric scientist William Gray of Colorado State University is pessimistic for the long term trend in storms: ". . . the U.S. has experienced a hurricane lull for the past twenty-five years. A correction is overdue, and when it comes, we're going to see hurricane damage like we've never seen before." Gray considers hurricanes to be "the biggest national threat facing the U.S."[61]

Hurricane *Andrew* seems to have become the measure of what the nation is faced with in tropical storms. *Andrew* was the alarm that woke the nation after a twenty-year lull in hurricane activity. During that twenty years, the population on the U.S. coasts increased by one third, putting millions more in the path of dangerous storms. Construction of homes, new industries, and agricultural crops put billions more in property dollars at risk. Weather also seems to be more violent. In the twenty-two years between 1947 and 1969, only one storm, *Camille*, was in Category 5, with winds over 155 MPH. In the twenty-two years between 1970 and 1992, two storms, *Hugo* in 1980 and *Andrew* in 1992, had winds from 131 to 155 MPH. In 1995 the coastal population from Texas to Maine was forty-four million, double the 1950 population. The question on everyone's mind is "When will the 'Big One' come?" With coastal evacuation time from twenty four to sixty-five hours, the question is critical.[62]

Hurricane recon crews are able to gather information that is much more accurate than weather satellites, according to Capt. Roy Deatherage, meteorologist of the 53rd WRS: "We can get it within a mile or two."[63] A storm of the magnitude of Category 5, at the top of the Saffer-Simpson scale, calls for the evacuation of all residents within ten miles of the coast. Minutes after *Andrew* made landfall, 750,000 homes lost power, roofs were peeled off, and hundreds of thousands of residents who chose not to evacuate huddled terrified in halls, closets, and bathrooms. The next morning, twenty-five thousand homes had been destroyed, leaving a quarter of a million people homeless, and one hundred thousand more homes were damaged. In four hours, *Andrew* had caused up to $30 billion in damages.[64]

The argument continues, with reason on each of the various sides. The real weakness of aerial reconnaissance and the value of an adequate number of back-up planes is that aircraft, being mechanical, often are out of commission when needed. Aircraft also grow older, and the "aging aircraft" problem is one that plagues the WC-130s of the reserve squadron. In Hurricane *Andrew*, for example, two missions were canceled because the planes were not flyable. A third aircraft was returning from Antigua with thirty maintenance people on board. The NHC sent it through the storm since there had been no fix for some time. Imagine thirty maintenance specialists flying through a storm! That's the sign of the aging aircraft problem we have," reported Dr. Sheets.[65]

The greatest single technical advancement, and the first choice of the National Hurricane Center in observing tropical meteorology, is the geosynchronous satellite. The history of satellite technology has been one of constant improvement. After the experimental Explorer II, October 1959, came TIROS-1, April 1960, which primarily provided cloud cover photography. NIMBUS-1, 28 August 1964, offered high resolution infrared radiometer for night pictures; and, on 10 September 1965, the Air Force Defense Meteorological Satellite program (DSMP) was launched. ESSA-1, 3 February 1966, became the first operational satellite with two wide-angle TV cameras. ESSA-2, launched twenty-five days later, providing automatic picture transmission (APT) of satellite data for meteorological analysis, forecasting, and the daily television pictures. The next advancement was the experimental Applications Technology Satellite (ATS) in geosynchronous earth orbit on 6 December 1966, with a "spin-scan" camera, invented by Verner E. Suomi of the University of Wisconsin. From this technology evolved the first Geostationary Operational Environmental Satellite (GOES) on 16 October 1975—producing images that span the entire hemisphere each thirty minutes. GOES remains the backbone of tropical meteorological observation and analysis today.[66]

Unfortunately, GOES has run out of fuel and remains in a stationary orbit on the western U.S. coast which restricts its use in the Atlantic. For observations of Atlantic storms, the NHC "borrows" time on European Meteosat-3, which gives a much better angle on the storms. On 25 February 1993, the European Satellite Meteosat-3 was repositioned at 75 degrees west, the normal position for the GOES-East, which allows coverage of the eastern Atlantic, important for hurricane detection and tracking.[67] New GOES satellites are being planned for the future to fill the gap between GOES and Meteorsat. Aircraft help fill in this gap at the present time (1995).[68]

Teaming with weather satellites, the double whammy to weather reconnaissance comes with advanced radar. From Congressional hearings came the prediction of the future: "The next generation weather radar, NEXRAD, as it was called in its development stage and as it is being implemented is the Weather Surveillance Radar 1988 Design Dopler, (WSR-88D), with twenty-one NEXRAD systems installed around the country. . . ."[69] And there is a new geosynchronous program—GOES-7.[70]

And so, airborne reconnaissance yields to satellites and new generations of radar positioned along the coast, and in other storm-prone parts of the nation. The greasy crew chief and the sweat-stained pilot and crew, in their olive drab flying machine, have given way to the satellite scientist, the radar screen observer, and the computer technician, in white shirt and proper tie, as they operate their orbiting "aircraft" invisible in the heavens, sending its life-saving messages.

Weather recon has traveled many miles. It is said that 144 crewmen died while the Air Weather Service ran weather reconnaissance, from 1942 until 1974. To their memory and to those unnumbered who died in weather missions before and after 1974, we dedicate this history.

Appendix: Chronology and Lineage of Weather Reconnaissance Squadrons

First Army Air Forces' Weather Reconnaissance Squadron,
Air Route Medium
1942-1945

Organized at Patterson Field, Ohio, 16 August 1942, as Army Air Force Weather Reconnaissance Squadron (Test), Number 1; sometimes called the 1st Weather Reconnaissance Squadron (Test)—Assigned to the 2nd Weather Region—Received its first aircraft, a C-45, on 16 August 1942—Moved to Truax Field, Wisconsin, 20 April 1943—Moved to Presque Isle, Maine, June 1943—Began operations in the North Atlantic in summer, 1943—Redesignated 30th Weather Reconnaissance Squadron 21 December 1943, with no change in station, and assigned to Headquarters Air Transport Command (ATC)—Winter 1943, reconnaissance flights were sent to cover South Atlantic ferrying routes from Morrison Field, Florida, to Puerto Rico to Trinidad and Brazil—Between 21 December 1943 and 21 August 1944, 30th Weather Reconnaissance Squadron Redesignated 30th Weather Reconnaissance Squadron, Air Route Medium-ATC—In August 1944, became the 1st Weather Reconnaissance Squadron, Air Route, Medium—ATC—On 5 and 6 September 1944, headquarters for the 1st moved from Presque Isle, Maine to Grenier Field, New Hampshire—With the end of the war, the squadron was inactivated on 21 December 1945.[1]

Second Weather Reconnaissance Squadron, Air Route Medium
1944-1945

Activated 1 February 1944 at Key Field, Meridian, Mississippi—Assigned to Third Reconnaissance Command, August 1944—Departed for the China-Burma-India Theater under Tenth Weather Region, Eastern Air Command—Arrived in India 14 October 1944, with eighteen B-25s and three P-38s—Inactivated 28 December 1945.

Special Army Air Forces Hurricane Reconnaissance Unit
1944

Activated 18 May 1944 by Joint Chiefs of Staff, Washington—Assigned to Army Air Forces Weather Wing, Asheville, North Carolina, 18 May 1944—Four crews with B-25D aircraft, divided into four flights—Flights 1 and 2 stationed at ATC's 36th Street Army Air Base (AAB), Miami, Florida; Flights 3 and 4 stationed at Bourenquen AAB, Puerto Rico—Flights 1 and 2 transferred to Morrison AAB, Florida, June 1—Unit deactivated 1 December and crews assigned to Tenth Weather Region, China-Burma-India Theater.[2]

655th-55th Weather Reconnaissance Squadron
1944-1975

Activated on 21 August 1944, at Will Rogers Field, Oklahoma—Assigned Twentieth Air Force, attached to XXI Bomber Command, 11 April 1945—Assigned Harmon Field, Guam, and Central Field, Iwo Jima, 11 April 1945—Redesignated 55th Reconnaissance Squadron, Long Range, Weather, 16 June 1945—Redesignated 55th Reconnaissance Squadron, Very Long Range, Weather, 27 November 1945—Assigned to ATC and, in turn, to Air Weather Service (AWS) 20 March 1946—Redesignated the 55th Strategic Reconnaissance Squadron, Medium, Weather, on 22 January 1951 and assigned to AWS 21 February 1951—Redesignated 55th Weather Reconnaissance Squadron 15 February 1954—Discontinued on 8 July 1961—Activated and assigned to Military Air Transport Service (MATS) 12 October 1961—Transferred to Aerospace Rescue and Recovery Service (ARRS) 1 September 1975.

53rd Weather Reconnaissance Squadron
1944-to date

Activated as 3rd Reconnaissance Squadron, Air Route, Medium, at Grenier Field, New Hampshire, 31 August 1944—Redesignated 3rd Reconnaissance Squadron, Weather, Heavy, 26 January 1945—Divided into three flights: Alpha, flying synoptic tracks over North Pacific, stationed at McCord Field, Washington; Bravo Flight, Gander Field, Newfoundland,

Goose Bay, Labrador, and Grenier Field, New Hampshire; and Flight Charlie operated from the Azores—Redesignated as 53rd Reconnaissance Squadron, Long Range, Heavy, Weather, 15 June 1945—Redesignated as 53rd Reconnaissance Squadron, Very Long Range, Weather, 13 March 1946—Moved to Morrison Field, Florida, 8 November 1946, flying missions in the North Atlantic—Deployed to Kindley Field, Bermuda, June and July 1947 and become a part of the 308th Reconnaissance Group, Weather, on 17 October 1946—All squadrons of the 308th were inactivated on 15th October 1947—373rd Reconnaissance Squadron, VLR, Weather, formed 16 September 1947, replacing the 53rd—Assigned to the 8th Weather Group, Fort Trotten, Long Island, New York—373rd inactivated on 19 February 1951 at Kindley Field, Bermuda—53rd Strategic Reconnaissance Squadron, Medium, Weather, activated on 20 February 1951 at the same base—53rd Strategic Reconnaissance Squadron, Medium, Weather, moved to Burtonwood, RAF Station, England, November 1953 with a flight in Dhahran, Saudi Arabia—Flight A of the 53rd remained at Kindley to accomplish hurricane and synoptic reconnaissance—February 1954, 53rd at Burtonwood and Flight A of 53rd at Kindley; redesignated 53rd Weather Reconnaissance Squadron—Flight A, 53rd Reconnaissance Squadron discontinued and 59th Weather Reconnaissance Flight activated at McKindley AFB, Bermuda on 8 May 1955, to assume mission of 53rd—59th Flight at Bermuda, Redesignated 59th Weather Reconnaissance Squadron 1 April 1956, at Bermuda—53rd Reconnaissance Squadron inactivated in England 18 March 1960—Reactivated on 8 January 1962 at Kindley AFB, Bermuda—Moved to Hunter AFB, Georgia on 8 September 1963—Assigned to the Ninth Weather Reconnaissance Wing, McClellan AFB, California—Moved to Ramey AFB, Puerto Rico, 15 June 1966, as a part of the Ninth Weather Reconnaissance Wing, McClellan AFB, California—Ramey closed and 53rd moved to Keesler AFB, Mississippi, 1 July 1973— Assigned to the 41st Rescue Weather Reconnaissance Wing of the ARRS of the Military Airlift Command (MAC) 1 September 1975—Seven WC-130s transferred to the 920th Weather Reconnaissance Group of the Air Force Reserve—920th *"Storm Trackers"* took over seventy percent of hurricane and east coast storm missions—Hurricane recon unit became known as Detachment 1, 7th Weather Wing and the 53rd WRS, AF Reserve, continued to fly all storm assignments—Previously discontinued, the 53rd Reconnaissance Squadron, Air Force Reserve, was officially reactivated 1 November 1993, with ceremonies January 1994 at Keesler Air Force Base, Mississippi, as the sole remaining unit of the once world-wide organization.

54th Weather Reconnaissance Squadron
1944-1975

Activated as 8th Reconnaissance Squadron, England, 23 March 1944—redesignated 654th Bombardment Squadron, Heavy (Reconnaissance Special), 17 July 1944—Redesignated 54th Reconnaissance Squadron, Long Range, Weather, 29 September 1945—Redesignated 54th Reconnaissance Squadron, VLR, Weather, 17 January 1946—Assigned to AWS March 1946—Moved to Langley Field, Virginia, June 1946—Moved to Morrison Field Florida, 21 July 1946—Assigned to 43rd Weather Wing, moved to Guam, 1 August 1947—Inactivated on 15 October 1947—Redesignated 54th Strategic Reconnaissance Squadron, Medium, Weather, 22 January 1951—Redesignated 54th Weather Reconnaissance Squadron 15 February 1954—Discontinued 18 March 1960—Reorganized 18 April 1962—Assigned to the 9th Weather Wing 8 July 1962—Assigned to the ARRS's 41st Rescue and Weather Reconnaissance Wing 1 September 1975.

56th Weather Reconnaissance Squadron
1951-1972

Constituted as 358th Fighter Squadron 12 November 1942—Inactivated 20 November 1946—Redesignated 56th Strategic Reconnaissance Squadron, Medium, Weather, 22 January 1951—Activated Misawa, Japan, 22 January 1951, assigned to 2143d Air Weather Wing 21 February 1951—Moved to Yokota AB, Japan, 14 September 1951—Redesignated 56th Weather Reconnaissance Squadron 15 February 1954—Assigned to 9th Weather Group 1 February 1960—Assigned to 9th Weather Reconnaissance Group 8 July 1961—Assigned to 9th Weather Reconnaissance Group 8 July 1962—Assigned 9th Weather Reconnaissance Wing 8 July 1965—Inactivated 15 January 1972.

57th Weather Reconnaissance Squadron
1954-1969

Constituted/activated 399th Fighter Squadron 1 August 1953, Hamilton Field, California—Redesignated 57th Weather Reconnaissance Squadron 11 July 1954—Inactivated 25 January 1956—Redesignated 57th Reconnaissance Squadron, Very Long Range, Weather, 3 July 1947—Inactivated 27 June 1949—Redesignated 57th Strategic Reconnaissance Squadron, Medium, Weather, 22 January 1951—Activated Hickam AFB, Hawaii 21 February 1951—Redesignated 57th Weather Reconnaissance Squadron 15 February 1954—Inactivated 18 October 1958—Activated and assigned to MATS 8 February 1962—Reorganized at Kirtland AFB, New Mexico and assigned to 9th Weather Reconnaissance Group 16 February 1962—Moved to Avalon AFB, Australia, 30 September 1962—Moved to Hickam AFB, Hawaii, 15 September 1965—Inactivated 30 November 1969.

58th Strategic Reconnaissance Squadron, Medium, Weather
1945-1974
Originally constituted/activated as 400th Fighter Squadron 1 August 1943 at Hamilton Field, California—Redesignated 58th Reconnaissance Squadron, Weather, 7 July 1945—Activated as 598th Weather Reconnaissance Squadron, Eielson AFB, Alaska, 20 February 1951—Redesignated 58th Strategic Reconnaissance Squadron, Medium, Weather, 22 January 1951—Activated at Eielson AFB, Alaska, 21 February 1951—Redesignated 58th Weather Reconnaissance Squadron, 15 February 1954—Last mission flown 30 June 1958 and inactivated on 8 August 1958. In May and June 1958, aircraft, crews, and most maintenance and administrative personnel moved to McCord AFB, Washington to form Detachment 3, 55th Weather Reconnaissance Squadron, effective 1 July 1958. On this date, many 58th maintenance and administrative personnel were transferred to Ladd AFB, Alaska, to form Detachment 1, 55th Weather Reconnaissance Squadron—Inactivated 8 August 1958—Reactivated 15 April 1963 at Kirtland AFB—Inactivated 30 June 1974.

59th Weather Reconnaissance Squadron
1945-1964
Constituted 59th Reconnaissance Squadron, Long Range Weather, 1 August 1945—Activated at Will Rogers Field, Oklahoma, assigned to III Reconnaissance Command 1, August 1945—Redesignated 59th Reconnaissance Squadron, Very Long Range, Weather, 27 November 1945—Assigned to ATC 13 March 1946—Assigned to the AWS 20 March 1946—Assigned to the 308th Reconnaissance Group, Weather, 17 October 1946—Moved to Ladd Field, Alaska, 1 June 1947—Inactivated 15 October 1947—Redesignated 59th Weather Reconnaissance Flight on 3 March 1955—Redesignated 59th Weather Reconnaissance Squadron 1 April 1956—Discontinued 18 March 1960—Reorganized at Goodfellow AFB, Texas, assigned to the 9th Weather Reconnaissance Group 8 July 1963—Discontinued and inactivated 8 May 1964.

373d Reconnaissance Squadron, Very Long Range, Weather
1947-1951
Constituted 373d Bombardment Squadron, Heavy, 28 January 1942—Activated at Gowen Field, Idaho, assigned to the 308th Bombardment Group 15 April 1942—Moved to Yankai, China 20 March 1943, and to Luliang, China 14 September 1944—inactivated 7 January 1946—Redesignated 373d Reconnaissance, Very Long Range, Weather, 16 September 1947—Inactivated on 19 February 1951.

374th Reconnaissance Squadron, Very Long Range, Weather
1947-1951

Constituted 375th Bombardment Squadron, Heavy, 28 January 1942—Moved to Chengkung, China 20 March 1943, to Hsinghing, China 18 February 1945 and to Rupsi, India 27 June 1945—Moved to Camp Kilmer, New Jersey and inactivated 6 January 1946—Redesignated 375th Reconnaissance Squadron, Very Long Range, Weather, 16 September 1947—Activated at Ladd Field, Alaska, assigned to 7th Weather Group 15 October 1947—Moved to Eielson AFB, Alaska 6 March 1949—Inactivated 21 February 1951.

375th Reconnaissance Squadron, Very Long Range, Weather
1947-1951

Constituted the 375th Bombardment Squadron, Heavy, 28 January 1942—Flew bombing missions in China—Inactivated 6 January 1946—Redesignated the 375th Reconnaissance Squadron, Very Long Range, Weather, 16 September 1947—Activated at Ladd Field, Alaska—One flight operated from Fairfield-Suisun AAF, California, and later from Shemya, AFB, Alaska, 15 October 1947 to 15 May 1949—Moved Eielson AFB, Alaska, 6 March 1949—Inactivated 21 February 1951.

512th Reconnaissance Squadron, Very Long Range, Weather
1947-1951

Constituted as the 512th Bombardment Squadron, Heavy, 19 October 1942—Flew bombing missions until redesignated 512th Reconnaissance Squadron, Very Long Range, Weather, 6 May 1947—Activated at Gravelly Point, Virginia, assigned to the 376th Reconnaissance Group, 27 May 1947—Assigned to AWS 16 September 1947—Inactivated 20 September 1948—Reactivated at Fairfield-Suisun AFB, California, assigned to the 30th Reconnaissance Group 13 February 1949—Assigned to 2143d Air Weather Wing, 14 November 1949 and moved to Yokota AB, Japan, 27 January 1050—Moved to Misawa, Japan, 11 August 1950—Inactivated 20 February 1951.

513th Reconnaissance Squadron, Very Long Range, Weather
1947-1951

Constituted 513th Bombardment Squadron, Heavy, 19 October 1942—Flew bombing missions until redesignated 513th Reconnaissance Squadron, Very Long Range, Weather, 6 May 1947—Activated at Gravelly Point, Virginia, assigned to the 376th Reconnaissance Group, 23 May 1947—Assigned to AWS 26 September 1947 and 308th Reconnaissance Group 14 October 1947—Inactivated 20th September 1948—Activated at Fairfield-Suisun AFB, California and assigned to 308th Reconnaissance Group 10,

August 1949—Detachment operated from Dhahran, Saudi Arabia, 6 March through May 1950—Assigned to AWS 19 December 1950—Inactivated 20 February 1951.

514th Reconnaissance Squadron, Very Long Range, Weather
1947-1951

Constituted the 514th Bombardment Squadron, Heavy, 19 October 1942—Flew bombing missions until 16 September 1947 when redesignated 514th Reconnaissance Squadron, Very Long Range, Weather, 16 September 1947—Activated at North Field, Guam—Assigned to the 43rd Weather Wing 15 October 1947—Inactivated 20 February 1951.

2078th Weather Reconnaissance Squadron, Special
1948-1950

Designated as the 1st Weather Reconnaissance Squadron, Special, 19 May 1948—Organized at Fairfield-Suisun AFB, California—Assigned to the 308th Reconnaissance Group, Weather 1 June 1948—Redesignated 2078th Air Weather Reconnaissance Squadron, Special, 1 October 1948—Discontinued 20 March 1950.

Notes

PREFACE

1. *Holy Bible*, Genesis 8:8-12, KJV.

2. Capt. M. L. Sevier, *Fact Book, Aerial Sampling and Weather Reconnaissance*, p. A-1-1.

3. John Bartlett, *Bartlett's Familiar Quotations*, p. 733b. This phrase is often attributed to Mark Twain, but his speech on weather in 1876, delivered to the New England Society, did not contain these words. However, *The Hartford Courant*, in an editorial presumably written by editor Charles Dudley Warner, did use the phrase.

4. Capt. Howard T. Orville, USN (Retired), "Stand By for Climate Control," *Saturday Evening Post*, November 29, 1958., p. 19.

5. Bill Nye, "Big Energy in Thin Air," *Omni*, March 1994, p. 8.

6. Ivan Ray Tannehill, *The Hurricane Hunters*, pp. 8, 9.

7. Sevier, *Fact Book*, p. A-1-2.

8. Bernard L. Wiggin, "A Weatherman has been called the most useful man of his age," *Smithsonian*, December 1970, pp. 50-51.

9. John Fuller, *Thor's Legions—Weather Support to the U.S. Air Force and Army, 1937-1987*, p. 21ff. Fuller has compiled one of the most complete and detailed histories of weather available, of which weather reconnaissance is a small part.

10. Eugene Linden, "Burned by Warming," *Time*, March 14, 1994, p. 79.

11. Vern Haugland, *The Eagle's War*, p. 133.

12. Capt. Walter Karig, Lt. Com. Russell L. Harris and Lt. Com. Frank A. Manson, *Battle Report,* pp. 90-104.

13. John Fuller, *Thor's Legions,* unpublished manuscript, p. 355.

14. Robert C. Sheets, "The National Hurricane Center—Past, Present and Future," *Weather and Forecasting,* June 1990, p. 200.

15. Wesley Frank Craven and James Lea Cate, *The Army Air Forces in World War II, Men & Planes,* Vol. 6, p. 617.

16. Fuller, *Thor's Legions,* unpublished manuscript, p. 243.

17. Special Orders No. 118, Army Air Forces Headquarters, Weather Wing, Asheville, N. C., 18 May 1944.

18. John K. Arnold, letter to the author, 28 December 1944.

19. Sevier, *"Fact Book,"p.* A-1-5.

20. David McCullough, *Truman,* p. 747.

21. Ibid., p. 749.

22. Craven and Cate, *The Army Air Forces in World War II,* p. 618.

23. Ibid., p. 311.

24. "Hurricane hunters add to jet power," *Greenville (Tex) Herald Banner,* 10 December 1994.

25. Maj. Douglas Lipscombe, *Hurricane Reconnaissance,* n.d, (c. 1994).

26. Robert Sheets, *NOAA'S Response to Weather Hazards—Has Nature Gone Mad?,* p. 37.

27. Virginia Woolf, *Mrs. Galloway,* 1925.

28. Dr. Frank Luther Mott, University of Missouri School of Journalism, 1948.

29. George Gissing, "Winter," *The Private Papers of Henry Ryecroft,* 1903.

PROLOGUE

1. SNAFU situation normal, all "fouled" up. This is a standard military phrase.

2. Detached service (DS)-when personnel were "loaned" from one military unit to another for specialized short-term duty. DS personnel usually enjoyed more freedom from military protocol than regulars, but were often overlooked for special awards, promotions, etc.

3. John J. Bergmeier, "Orphans of the Storm," *Air Classics,* April 1986, p. 29.

4. "Weather and Mediums," *History of the Ninth Bomber Command,* p. 14. Cecil Dotson, combat weather officer, Ninth Bomber Command, correspondence with the author, 25 May 1994.

5. Jim Swarts, "Orphans of the Storm," *Stars and Stripes,* 12 October 1944, p. 21.

6. Fuller, *Thor's Legions,* p. 79.

7. Swarts, "Orphans of the Storm," p. 21.

8. Bergmeier, "Orphans of the Storm," p. 32.

9. Ibid.

10. "G" means gravity. One G is the normal pull of gravity, such as when flying straight and level. In a tight turn the pilot may experience the pull of two to five Gs. In violent maneuvers a pilot may exceed more Gs than the body can stand and blacks out when blood is forced from the head to the lower body. Normally a pilot can endure three to five Gs before blackout.

11. Burgmeier, "Orphans of the Storm," p. 57.

12. Ibid., p. 74. These two articles contain the most complete information about the Ninth Weather Reconnaissance Squadron (Provisional) as was found in any military history. They were written in memory of the author's father, Lt. Col. John J. Burgmeier, Jr., who flew with the "Orphans" as Flight Officer Burgmeier in his P-51D *St. Cloud,* named after his hometown in Minnesota. He retired in 1974.

13. Kent Zimmerman, unpublished *Personal Memoirs,* "Air Weather Service," Chapt. 18, p. 204.

14. A small tractor used to tow aircraft.

15. Aubrey O. Cookman, "Top of the World Weather Run," *Popular Mechanics,* November 1948, pp. 99-100.

16. Charles Monasee, letter to the author 2 June 1994; John Gilchrist, letter to the author, 20 May 1994.

17. John Bortz, navigator across the North Atlantic, 1943.

18. Cookman, "Top of the World Weather Run," p. 262.

19. Lewis Carroll, *The Walrus and the Carpenter,* st. 1, lb.

20. Edward R. Murrow, *In Search of Light,* p. 263; Alexander Kendrick, *Prime Time: The Life of Edward R. Murrow,* p. 344.

21. Patrick Hughes, *A Century of Weather Service,* p. 115.

22. Personal experience of Capt. Robert E. Guthland, Warrensburg, Missouri, retired air force meteorologist.

23. "Greenhouse Fact Sheet," Defense Nuclear Agency, Washington, D.C. n.d. (c. 1951).

24. A "rem" is a unit of radiation exposure with a biological effect equal to one roentgen of x-ray. Rem is an acronym of **r** (roentgen) **e** (equivalent in) **m** (man), *The Random House Collegiate Dictionary, Revised Edition,* 1988, p. 1115.

25. Eugene Wernette, personal letter to the author, 11 April 1994.

1. Dwight Eisenhower, *Crusade in Europe,* 246.

2. Bruce W. Nelan, "Ike's Invasion," *Time,* June 6, 1944, p. 40.

3. The term "D-Day" is used to denote the date for any planned military operation. Since *Operation Overlord* was such a significant event, D-

Day has become, almost exclusively, the accepted name for the Normandy invasion.

4. Cecil Dotson, "How D-Day Was Chosen," *Air Weather Reconnaissance Newsletter, D-Day Commemorative Edition,* June 1994, pp. 2-7.

5. Eisenhower, *Crusade in Europe,* p. 239.

6. Robert C. Bundgaard, "D-Day Forecast, Fifty Years Later," *Air Weather Association Newsletter, D-Day Commemorative Edition,* June 1994, p. 7.

7. Irving P. Krick with Lee Edson, "How D-Day Was Chosen," *Newsletter,* Air Weather Association, June 1994, p. 2.

8. Cecil Dotson, "D-Day and the Weather, *Air Weather Association Newsletter, D-Day Commemorative Edition,* June 1994, p. 4.

9. Ibid.

10. John Fuller, unpublished manuscript for *Thor's Legions,* p. 242.

11. Krick and Edson, "How D-Day was Chosen," p. 7.

12. Ibid., p. 2.

13. Bundgaard, "D-Day Forecast, Fifty Years Later."

14. John Fuller, *Thor's Legions,* p. 18N.

15. Patton, *Lend-Lease,* p. 169.

16. Art Gulliver, "Langford Lodge," *Newsletter,* Air Weather Reconnaissance Association, 14 December 1993. p. 4.

17. Fuller, *Thor's Legions,* p. 58.

18. Bob Bundgaard and Paul Gross, "Yates' Necrology," *Bulletin of the American Meteorological Society,* June 1944.

19. Ibid.

20. Jack C. Calloway, personal log sent to the author, 17 February 1993.

21. Fuller, p. 18.

22. William Young, "The Meteorological Air Observer's Standpoint," *Air Weather Reconnaissance Newsletter,* Vol. 1, July 1998, p. 4.

23. Fuller, *Thor's Legions,* unpublished manuscript, p. 189.

24. Bundgaard and Gross, "Yates' Necrology".

25. Swarts, Historical Division Air Weather Service, "Tracing of Organizational Structures, Weather Reconnaissance Squadrons, End of W.W.II to Present." 13 February 1959, pp. 1-2.

26. "Weather," *AAF, The Official World War II Guide to the Army Air Forces,* p. 238.

27. W. F. Craven and J. L. Cate, *The Army Air Forces in World War II,* Vol. VII, p. 311.

28. P. G. Rackliff, "Meteorological Reconnaissance Flights," *Meteorology and World War II,* p. 1.

29. Ibid., p. 2.

30. Cathy Willis, "Rich" Courtney, Frank Baillie, "Aerographers Mates

Celebrate," Brochure of the Annual Reunion of the National Weather Service Association, 1993, p. 10.

31. The Navajo Indian language was one of the most successful codes used during W.W.II. The Japanese were never able to break it.

32. James F. Van Dyne, personal letter to the author, 21 November 1994.

33. L. D. Clark, personal letter to the author, 20 January 1995.

34. Ibid.

35. *Summary of 17th Weather Squadron, World War II History*, microfilm from the Air Force History Section, Maxwell Field, Alabama. Submitted by James F. Van Dyne, in a personal letter to the author, 28 September 1944.

36. Arthur McCartan, letter to the author, 6 February 1994.

37. Ibid.

38. One of the horrors of war is that, statistically, a certain number of aircrews are doomed to death before any operation. Commanders weigh these odds as campaigns are planned.

39. John Fuller, "A Pioneer in Weather Recce," *Air Weather Service Observer*, March 1985, p. 3.

40. L. S. Jones, Edward J. Machala, *History of AAF Weather Reconnaissance Squadron (Test) No. 1*, 21 August 1942-15 December 1943, p. 4.

41. Charles Bates, letter to the author, 5 July 1994. Bates was a weather forecaster along the North Atlantic in 1943 and author, with John Fuller, of *America's Weather Warriors, 1814-1985*.

42. Ibid.

43. Ibid.

44. Fuller, *Thor's Legions*, p. 63.

45. L. S Jones and Edward J. Machala, *History of AAF Weather Reconnaissance Squadron, Test No. 1, 21 August-15 December 1943.*, p. 18, 36.

46. Bates, letter.

47. "United National Air Forces Flight Forecast", prepared by Charles Bates, weather forecaster; with flight log Stephenville to Prestwick, by Jack Catchings, senior American Airlines pilot, December 2, 3, 1943.

48. A flight simulator used to train pilots to fly instruments and to navigate from station to station-all safely on the ground. The Link was a wonderful training aid, but it did not offer real weather conditions to allow pilots to build confidence.

49. TDY-Temporary Duty.

50. Kilroy was a famous nonentity who was everywhere the Air Forces flew. I saw "Kilroy was here," from the world's most horrible men's room in Petionville, Haiti, to the ceiling of a hotel room in Darjeeling, India. I knew Kilroy had been there and all was well.

51. Clark Hosmer, personal letter to the author, 15 July 1994.

52. For a beautifully written description of approaching Greenland and flying into BW-8, see Ernest K. Gann, *Fate is the Hunter,* Simon and Schuster, 1961.

53. Guy Murchie, *Song of the Sky,* p. 233.

54. In research for this history, the source of this story was found in Fuller, p. 60.

55. Personal experience of the author, June 1944.

56. Norman Carlisle, "Our Weather is Changing," *Coronet,* January 1954, pp. 17ff.

57. Jones and Machala, *"History of AAF Hurricane Reconnaissance Squasron, Test 1 . . ."* Vol. II, p. 5.

58. Personal experience of the author.

59. James A Johnson, personal letter to the author, 25 January 1994.

60. Ibid.

61. Weather letdown procedures along the North Atlantic, 1944-45.

62. Before pilots are allowed to take off, the base operations officer must sign the flight clearance, thus assuming responsibility for the flight meeting all safety and weather requirements. Pilots, under some conditions, were allowed to sign their own clearance which relieved base operations of responsibility.

63. Glenn Bradbury, personal letter to the author, July 29 1994.

64. Pressure pattern flying was popular in the early days of trans-oceanic flight. Winds circulate clockwise around a high pressure system and counter-clockwise around a low pressure system. By flying (westward) across the top of a low pressure system or across the bottom of a high pressure system, the pilot can take advantage of tail winds. On eastbound flights, the opposite is true.

65. Fuller, *Thor,* p. 62.

66. Craven and Cate, *"The Army Air Forces in World War II,"* p.320.

67. Ibid., p. 323.

68. Kent Zimmerman, unpublished personal memoirs, p. 194.

69. Clark Hosmer, personal letter to the author, July 25, 1994.

70. Fuller, *Thor's Legions,* p. 64. Arthur McCartan, personal letter to the author, 23 March 1994.

CHAPTER 2

1. Andrew H. Brown, "Men Against the Hurricane," *The National Geographic Magazine,"*October 1950, p. 537; Guy Murchie, *Song of the Sky,* p. 136.

2. Joseph N. Kane, *Famous First Facts,* 3rd ed., p. 306.

3. Murchie, *Song of the Sky,"*p. 137.

4. Brown, "Men Against the Hurricane," p. 537.

5. M. L. Sevier, *Aerial Sampling and Weather Reconnaissance*, November 1983, p. A-1-5.

6. Names were not given to tropical cyclones at that time.

7. "Contact" means that the pilot is able to fly and navigate by reference the ground.

8. Joseph B. Duckworth, *Flight Through a Tropical Hurricane*, August 18, 1943.

9. Duckworth, personal letter to F. W. Reichelderfer, Chief of the U.S. Weather Bureau, 13 September 1943.

10. Sevier, *Aerial Sampling and Weather Reconnaissance*, p. A-1-5.

11. Wesley Frank Craven, James Lea Cate, *The Army Air Forces in World War II, Men and Planes*, Vol. 6, p. 360.

12. Albert W. Friday, Jr., testimony before the Subcommittee on Space, Committee on Science, Space and Technology, U.S. House of Representatives, 14 September 1993, p. 19. Mr. Friday is Assist. Administrator for Weather Service, National Oceanic and Atmospheric Administration, U.S. Department of Commerce.

13. "Storm Cripples South U.S. Airline, Military Facilities," *Aviation Week & Space Technology*, 31 August 1992, p. 27.

14. Friday, "Testimony Before the Subcommittee on Space," p. 19.

15. Lt. Col. Gale Carter, personal letter to the author, 10 March 1994.

16. Friday, "Testimony Before the Subcommittee on Space," p. 19.

17. Brown, "Men Against the Hurricane," pp. 537-538. Major George Ashley, "1944—Year of the Great Hurricane," *Science Digest*, May 1944, pp. 39-40.

18. Ivan Ray Tannehill, *The Hurricane Hunters*, pp. 70-71.

19. Ibid., pp. 71-72.

20. Ibid., p. 75.

21. John Fuller, *Thor's Legions*, p. 242n.

22. James B. Baker, telephone conversation with the author, 6 March 1995.

23. Gordon H. MacDougall, "August 1943, Flying into a Hurricane," *Weatherwise*, August 1986, p. 216ff; MacDougall, personal correspondence with the author from 12 July, 15 August 1994; Ivan Ray Tannehill, *Hurricanes of the Twentieth Century*, p. 231-232.

24. Conversations with pilots from Caribbean Air Command, 1944.

25. Gary F. Frey, "The Hurricane Hunters," *Journal American Aviation Historical Society*, Spring 1980, p. 46.

26. Personal experience of the author, April 1944, Columbia Army Air Base, Columbia, S. C.

27. AC/AS OC&R-AFRWX, Major D. J. Smith 74727, 26 May 1944, a copy in the author's files. In the Chief of Staff directive, Captain Wiggins'

first name is spelled "Allan." On all subsequent orders his name is correctly spelled "Allen," with the same serial number. "

28. Special Orders No. 112, Hq. Columbia Army Air Base, 21 April 1944, By order of Colonel McNeil.

29. Ibid.

30. H. C. Sumner, "North Atlantic Hurricanes and Tropical Disturbances of 1944," *Monthly Weather Review*, December 1944, p. 340.

31. Statement from Fred Huber, San Antonio, Texas, 10 December 1994.

32. Milton Sosin, "Daily News Writer Hunts Hurricane in 100 MPH Winds," *Miami Daily News*, 13 September 1944.

33. *Life*, "1944 Hurricane," 2 October 1944, p. 41.

34. *Time*, "Disasters, The Great Whirlwind," 25 September 1944, p. 14-15.

35. Tape interview with Fred Huber, 15 January 1995.

36. Memo 201.22, "Assignment of Personnel," to Commanding Officer, AAF Weather Wing, Asheville, N. C., from Col. John K. Arnold, Regional Control Officer, Ninth Weather Region, West Palm Beach, Florida, 28 December 1944.

37. Letter, Jesse Jones to the Secretary of War, Washington, D.C.; Letter from Henry L. Stimson, to the Secretary of Commerce, Washington, D. C.; Letter from F. W. Reichelderfer, Chief, Weather Bureau, U. S. Department of Commerce, to Commanding General, Weather Division, Headquarters, Army Air Forces, Arlington, Virginia; all letters dated 28 September 1944.

38. Murchie, *Song of the Sky*, pp. 137-138.

39. Personal experience of the author with his weather officer Lt. George Abersold, 1944.

40. Murchie, *Song of the Sky*, p.138.

41. Capt. L. Scott Cox, "Hunting for Trouble," *The Mac Flyer*, January 1978, p. 13ff.

42. Personal experience of the author, September 1944.

43. William Anderson, *Hurricane Hunters*, p. 85-86. While this book is fiction, it is based on fact and technical details are as accurate as any history. Anderson was a hurricane pilot with two thousand hours of weather recon experience.

44. Tom Robison, letter to the author, January 1994.

45. Andrew Sparks, "Hurricane Hunters of Hunter," *The Atlanta Journal and Constitution Magazine*, 30 August 1964, p. 3.

46. U.S. Department of Commerce, "The Naming of Hurricanes," undated.

47. U. S. Department of Commerce, "Notes on the Naming of Hurricanes-1956," 23 December 1955, p. 2.

48. Kappler's Hurricane was named before the formal naming procedure was adopted, which prohibited specific individual's names being used.

49. Tannehill, *The Hurricane Hunters,* pp. 132ff.

50. John Fuller, unpublished manuscript of *Thor's Legions,* p. 443.

51. Don Offerman, personal letter to the author, 21 January 1994.

52. David Magilavy, personal letter to the author 29 June 1993.

53. Maj. Jim Perkins, "Documentary on the 53rd Weather Reconnaissance Squadron Hurricane Hunters," script for a television production by the same name, produced by Lt. Col. Paul McVickar, undated, (c. 1990).

54. Ibid.

55. Alwyn T. Lloyd, "B-29 Superfortress," *Detail and Scale,* Vol. 25, 1987, p. 57.

56. Perkins, "Documentary on the 53rd Weather Reconnaissance Squadron Hurricane Hunters," statement by Robert Sheets, director of the National Hurricane Center, p. 13.

CHAPTER 3

1. Claire Lee Chennault was a general in the Chinese Air Force, and the crews of the American Volunteer Group—The Flying Tigers—were civilians. Chennault rejoined the Army Air Force in March 1942 and was given the rank of brigadier general.

2. Harry M. Albaugh, personal letter to the author, 22 June 1994.

3. Arthur McCartan, personal letter to the author, 23 March 1994.

4. Eugene Wernette, personal letter to the author, 11 April 1994.

5. Charles Markham, "A Balloon Blower in Wartime China," publication unknown, date unk, (c. 1994) p. 23.

6. Bob Klossner, letter to the author including excepts from his personal memoirs, 17 February 1993.

7. Ivan Ray Tannehill, *Weather Around the World,* pp. 13, 113.

8. Ibid., p. 51.

9. Montgomery Truss, personal letter to the author, 4 March 1994.

10. Ibid.

11. William Tunner, *Over the Hump,* p. 63.

12. John Fuller, *Thor's Legions,* p. 110.

13. W. F. Craven and J. L. Cate, *The Army Air Forces in World War II,* Vol. 7, p. xxi.

14. *History of the Tenth Weather Squadron,* June 1942, Sept. 1945, p. 7.

15. Air Transport Command, *A History of Hump Operations,* Microfilm, 1 January-31 March 1945.

16. "Richard E. Ellsworth, 1911-1953," necrology of Ellsworth, written by his West Point classmates, and read on the occasion of the dedication of Ellsworth Air Force Base, Rapid City, South Dakota, by President Eisenhower, June 1953.

17. Microfilm, Ibid., p. 8-9.

18. Tunner,*Over the Hump*, p. 63. The author went to the CBI in March 1945, just after the "no weather" edict had been lifted. General Tunner had rescinded the rule, but because it prevented many canceled flights and moved freight over the Hump, it was not officially lifted until after the Big Storm of January 1945, by orders from Washington. See *Newsletter*, Hump Pilots Association, Summer 1987, p. 26.

19. *History of the Tenth Weather Squadron*, Ibid. p. 2.

20. Ibid. p. 7. For a dramatic account of building the airfields and the B-29 mission in China, see Spencer, *Flying the Hump*, p. 107ff.

21. Otha C. Spencer, *Flying the Hump*, p. 110.

22. Winton R. Close, "B-29s in the CBI-A Pilot's Account," *Aerospace Historian*, March 1983, p. 12.

23. Ibid., p. 8.

24. Fuller, *Thor*, p. 117.

25. The C-109 was a tanker version of the B-24. It's nickname, given by pilots who flew gasoline over the Hump, was "C-1-0-Boom!"

26. Rudyard Kipling, *The Naulahka*, ch. 5.

27. Telephone conversation with Harry Albaugh, 15 February 1995.

28. During World War II, aircraft required 100-octane fuel for highest performance. Often 90-octane was the only fuel available. This was usually satisfactory for cruise flight, but dangerous for take-off, climbing and landing. Modern automobiles, today, operate on 87-octane or better.

29. An indictor to show wind direction on the ground. It is normally called a "wind sock."

30. Harry Albaugh, telephone conversation with the author, 6 March 1995.

31. *History of the Tenth Weather Squadron*, Ibid.

32. Fuller, *Thor's Legions*, p. 118.

33. Charles H. Broman, letter to the author, 4 October 1994.

34. Personal experience of the author in China, 1945.

35. Robert Blake, B-25 navigator, from Lubbock, Texas, tells a tragic story: "It was 1:30 p.m. on 22 March 1945, as Capt. William "Tex" Waggaman, San Antonio, Texas, and I were on special duty at Hsian, China. On that day our mission was low-level bombing to destroy Japanese railroads and bridges. In the ten days we had lost three B-25s. Waggaman would be pilot and I would be navigator on what would be the final flight of our last B-25H.

"A Japanese supply train had been sighted about two hundred miles away. . . . Our job was to destroy the supply train with the 75mm cannon in the B-25H . . . I not only navigated but also loaded the 75mm cannon—the B-25H had no copilot. Captain Waggaman made a pass at the train and bullets ripped through our nose and fuselage. . . . I warned him to stay away from some buildings where heavy small-arms fire was coming from. . . . Waggaman made a second, slower approach straight for the buildings and fired the cannon at eight hundred feet. Just as he did, he made a convulsive lunge forward, blood streaming down his face. . . . He died instantly.

"I grabbed the wheel and leveled the plane. . . . A few hours of pilot training had taught me that much. . . . Other crewmen tried to remove Waggaman's body, [but] we weren't high enough to bail out, and no one had their 'chutes on. So we decided to try to get the plane back to Hsian, and, since I dared not try to land the plane, everyone would bail out. . . . Sgt. James Anderson got Waggaman's body out of the seat, although his foot was jammed between the seat and the controls. . . . We climbed to ten thousand feet to clear the mountains and, adding 180° to the course we had flown out, I knew we would come somewhere near the base. . . .

We passed over Hsian airfield in bright moonlight, about 8:00 p.m. and flew about ten miles west into more friendly Chinese territory and dropped the pilot's body and the others bailed out. I tried desperately to hold the plane level, faced the bottom hatch, turned loose and plunged out, . . . the 'chute opened as the plane crashed about half a mile away.

"I walked toward the burning wreck and . . . was rescued by Chinese guerrilla fighters. We spent the night in a cave. . . . Next morning, an Army jeep came into their camp . . . they had found me fifty-five miles west of Hsian through the Chinese 'grapevine.' Sergeant Anderson had ridden a donkey to the base and the Chinese had found Captain Waggaman's body and brought it in that morning. . . . He was buried 'somewhere in China.'" Later, Waggaman's body was moved and reburied in Austin, Texas.

36. Spencer, *Flying the Hump*, p. 116.

37. Stanford Kahn, personal letter to the author, 16 February 1994.

38. Ibid.

39. Howard Lysaker, person letter to the author, 12 July 1994.

40. Craven and Cate, *The Army Air Forces in World War II*, Vol. 5, p. 68.

41. Enzo Angelucci, Paolo Matricardi, *World War II Airplanes*, Vol. 2, p. 79-80.

42. Howard Lysaker, letter to the author, 12 July and 30 August 1994.

43. Robert Lashbrook, letter to the author, 17 November 1994.

44. Guy Murchie, *Song of the Sky*, New Revised Edition, p. 218; *Encyclopedia Britannica*, Vol. 19, 1954, p. 828.

45. Kent Zimmerman, *Memoirs of World War II*, p. 220; Personal letter to the author, 21 March 1994.

46. Craven and Cate, *The Army Air Forces in World War II*, Vol. 5, p. 576.

47. Murchie, *Song of the Sky*, p. 157.

48. Ibid, p. 265.

49. Elmar R. Reiter, *Jet Stream Meteorology*, p. 3ff.

50. Ibid., p. 8.

51. Murchie, *Song of the Sky*, p. 158-159.

52. Personal experience of the author and Fred Huber. Huber has retired to San Antonio, Texas.

CHAPTER 4

1. Barbara Tuchman, *Stilwell and the American Experience in China, 1911-1945*, p. 522.

2. Inscription on the 20th Air Force wall, Air Force Academy, Colorado, dedicated 9 June 1944.

3. Lewis L. Howes, *The Nuclear Explosion Detection System and the Role of the Air Weather Reconnaissance Program*, n.d. (c.1994), p. 21.

4. Tim Frable, "The End of Possum Red," tape transcription sent to the author.

5. News release, public relations officer, CQ, 20 AAF, CSC, Iwo, to COMGEN, 10 AAF, Guam, 15 August 1945.

6. John Hug, taped account of his weather recon experiences, sent to the author, 4 April 1994.

7. Historical Division, Air Weather Service, *Special Historical Study*, 13 February 1959.

8. The most difficult part of this entire history has been trying to untangle the maze of dates, name changes, activations and de-activations, etc. Thus the emphasis will be on missions and the work that the squadrons and their air crews performed in weather reconnaissance and, at the same time trying, to determine a consensus of chronological organization.

9. Eric Larraby, *Commander in Chief, Franklin Delano Roosevelt, His Lieutenants and Their War*, p. 580.

10. Craven and Cate, *The Army Air Forces in World War II, Men and Planes*. Vol. 6, p. 360.

11. The Boeing Company, "Boeing B-29 Superfortress," 13 January 1971.

12. Capt. Lyle P. Earney, "At War and Peace, the B-29 Superfort Served Well," *Tropic Topics*, (Guam), 25 January 1957.

13. Eugene Wernette, letter to the author, 11 April 1994; conversation with David Magilavy, 28 October 1994.

14. Edward Vercelli, letter to the author, 10 April 1994.

15. Howes, *The Nuclear Explosion Detection System*, p. 23.

16. Dr. Robert Sheets, director of the National Hurricane Center, *Documentary on the 53rd Weather Reconnaissance Squadron "The Hurricane Hunters."* Script for "History of the Hurricane Hunters," n.d. (c. 1992), p. 13.

17. Ray Wagner, *America's Combat Planes*, p. 3.

18. John Fuller, unpublished manuscript, *Thor's Legions*, p. 432.

19. Fuller, *Thor's Legions*, . . . , p. 242n.

20. Neal M. Bowers, "Bikini Atoll," *The World Book Encyclopedia*, 1960, p. 231.

21. "Weather and the Atomic Bomb Tests at Bikini," American Meteorological Service *Bulletin*, No. 8, October 1946, p. 435-37.

22. Charles Monasee, letter to the author, 2 June 1994.

23. Correspondence with David Magilavy who conferred with Charles Monasee about the details of these flights.

24. In atmospheric radiation sampling, air crews referred to drifting radioactive gases as a "hot cloud."

25. David McCullough, *Truman*, p. 747.

26. Ibid., p. 749.

27. Fuller, *Thor's Legions*, p. 247.

28. Howes, *Nuclear Explosion Detection System*, p. 29.

29. David Magilavy, letter to the author, 25 June 1994; interview with Jack Shelton, 12 March 1994.

30. Howes, *Nuclear Explosion Detection System*, pp. 15-16.

31. Alwyn T. Lloyd, *B-29 Superfortress in Detail & Scale*, Vol. 25, 7 July 1987, pp. 55-56.

32. Tom Robison, *Brief Chronology of Air Forces Aerial Weather Reconnaissance*, October 1994, p. 2.

33. Ibid.

34. Magilavy, letter.

35. Lester Ferriss, letter to Lewis L. Howes, 14 November 1994.

36. Ibid.

37. Howes, *Nuclear Explosion Detection System*, p. 18.

38. S/Sgt. John Spade, "Petrel Mission," *Trans-Pacifican*, 5 September 1952.

39. Ibid.

40. David Magilavy, letter to the author, 29 January 1993.

41. Lloyd, *B-29 Superfortresses in Detail & Scale*, p. 57.

42. "Air Weather Service Reconnaissance Tracks," AWSR, 55-12, 1

43. Aubrey O. Cookman, "Top of the World Weather Run," *Popular Mechanics*, November 1948, p. 97.

44. "Weather Plane Makes Long Flight Over Polar Region," *Vallejo* (California) *Times-Herald*, 23 October 1947, p. 11.

45. Leaflet, with photograph, from the files of Lorin Childers, Lacey, Washington.

46. John Matt, *Crewdog*, pp. 282-284.

47. Monasee, letter.

48. John Hug, transcript from taped interview, 4 April 1944.

49. John Gilchrist, letter to the author, 20 May 1994.

50. "Ice Islands as Bases," *Science News Letter*, 25 November 1950, p. 58; "Ice Islands," *Time*, 27 November 1950, p. 66.

51. *Science News Letter*, "Ice Islands as Bases."

52. Edward Phillips, letter to the author, 24 July 1995, excerpts from "Memories of a Maintenance Man," personal memoir.

53. Cathy Willis, "Rich" Courtney, Frank Baillie, "Aerographers Mates Celebrate," a history of Naval Weather Service, written for the 69th reunion of the National Naval Weather Service Association, 1993, p. 33.

54. Patrick Hughes, *A Century of Weather Service*," p. 101; Willis, Courtney, Baillie, "Aerographers Mates Celebrate.

55. Arthur McCartan, letter to the author, 6 February 1994.

56. Ibid.

57. Hug, tape transcript.

58. Ibid.

CHAPTER 5

1. John Fuller, *Thor's Legions*, p. 255

2. Capt. G. H. Behrens, "Trouble Shooters," *Pacific Patrol*, March 1952, p. 25.

3. Fuller, *Thor's Legions*, p. 266.

4. Ibid.

5. Alwyn T. Lloyd, "The B-29 Superfortress," *Detail & Scale*, p. 57.

6. Ibid.

7. Eugene Murphy, letter to the author, 29 September 1994.

8. Tom Robison, *Brief Chronology of Air Force Weather Reconnaissance*, p.1.

9. Fuller, *Thor's Legions*, p. 243.

10. Leon Attarian, letter to the author, 20 February 1995.

11. Cover your ass.

12. Bob Mann, "The Saga of the Stolen Engines," *Typhoon Topics*, September 1994.

13. Interview with Gen. Robert Moeller, USAF Retired, 29 October 1994, at Tucson, Arizona.

14. Marshall Balfe, letter to the author, 25 November 1994.

15. Charles Markham, *Granpap's Tracks Through Time*, personal memoirs undated.

16. Ibid.

17. *Richard E. Ellsworth, 1911-1953*, a tribute written by members of his West Point class of 1935. A copy of this tribute was sent to the author by Tex Albaugh, Mariposa, California.

18. Col. William C. Anderson, "Edward R. Murrow and the Eye of the Hurricane," *The Retired Officer Magazine*, April 1955, p. 55ff.

19. Edward Bliss, Jr., ed. *In Search of Light*, p. 263.

20. Ibid., p. 264.

21. William Anderson, "Edward R. Murrow and the Eye of the Hurricane," *The Retired Officer Magazine*, April 1995, pp. 55ff.

22. Robert Simpson,*"Documentary on the 53rd Weather Reconnaissance Squadron, 'The Hurricane Hunters'* undated, (c. 1990), p. 20.

23. Lt. Herbert G. Dorsey, Jr., and Lt. Oscar Shaftel, "Sky Quirks," *Air Force*, June 1994, pp. 31-32.

24. Arthur McCartan, letter to the author, 6 February 1994.

25. Glenn Bradbury, "AWS Conducts Aerial Radiological Surveillance," July 1994.

26. David Magilavy, letter to the author, 1 February 1994.

27. Ibid.

28. Lester Ferriss, letter to the author, 14 November 1994.

29. David Magilavy, *Air Weather Reconnaissance Newsletter*, Vol. 1, July 1993, quoting from a letter from the Program Manager, Nuclear Test Review, defense Nuclear Agency, Alexandria, Virginia, 2 June 1993.

30. *Newsletter, Pacific Air Weather Squadrons*, February 1995.

31. Donald Schertz, letter to the author, 22 November 1994.

32. DR—dead reckoning—navigation by time, ground speed and heading from a known position. It is less precise than radio navigation because of unknown or inaccurate factors such as wind speed and direction, slight changes in headings, and variations in aircraft speed.

33. An emergency speed when the engine manifold pressure and propeller rpm are set for absolute minimum fuel consumption and the aircraft trimmed for best flight attitude.

34. "Well Done," a tribute to Lt. Chester M. Stone and Lt. Vernon N. Barnett, *Flying Safety*, Hq. United States Air Force, May 1950, p. 21.

35. John Hug, transcript from taped interview, 4 April 1994.

36. "Eleven Crew Members Perished in B-29 Crash—Two Survive," *Bermuda Skyliner*, 9 November 1949.

37. Hug tape.

38. John A. Cox, "Hurricane Hunters," *Bee-Hive, United Aircraft Corp.*, 1959, p. 200; quoted by Gary F. Frey, *Journal*, American Meteorological Society, Spring 1980, p. 48.

39. Robert Simpson, *Documentary on the 53rd Weather Reconnaissance Squadron, "The Hurricane Hunters,"* undated, (circa 1990), p. 16.

40. Donald H. Offerman, letter to the author, 21 January 1994.

41. Ibid.

42. Simpson, *Documentary on the 53rd Weather Reconnaissance Squadron*, p. 14.

43. Eugene Wernette, letter to the author, 11 April 1994.

44. Ralph N. Phillips, "Navigation Problems in England," *Air Force,* June 1944, p. 36.

45. Lloyd, Ibid., pp. 58-59.

46. John Gilchrist, letter to the author, 12 May 1994.

47. David G. McFarland, "A Cold War Weather Mission," *Weatherwise,* n.d., p. 70.

48. Ted C. Jafferis, letter to the author, 1 July 1994.

49. Conversations with former weather reconnaissance crew members at a reunion of the Air Weather Association, Tuscon, Arizona., 27-30 October 1994.

50. Bradbury, "AWS Conducts Aerial Radiological Surveillance."

51. Ibid.

CHAPTER 6

1. James R. Slaeker, letter to the author, December 1994.

2. David Magilavy, letter to the author, 1 February 1994.

3. David G. McFarland, letter to the author, 26 September 1994.

4. Leon Attarian, letter to the author, 20 February 1995.

5. Slaeker, letter.

6. Magilavy, letter.

7. Ibid.

8. Ibid.

9. John Field, letter to the author, 27 November 1993.

10. A stable aircraft is designed, with hands-off flight, to seek straight and level flight attitude. It constantly tries to stabilize itself from outside forces such as turbulence, and is easier to fly.

11. John F. Fuller, "A Lesson from History . . . the WB-50D," *United States Air Force Friends Bulletin,* Summer/Fall 1986, p. 18.

12. News release, 56th WRS, Yokota, Japan, 16 September 1965.

13. John Fuller, "Brief History of 55th Weather Reconnaissance Squadron, 1944-1962," p. 5.

14. Ibid.

15. Fuller, "A Lesson from History . . . ," 5.

16. David Magilavy, fax to the author, 29 June 1994.

17. John D. Gilchrist, letter to the author, 20 May 1994.

18. George Horn, "Final Disposition of WB-50Ds," letter to the author, 10 March 1994.

19. Capt. Morris Gibson, Jr., "Overview of Aerial Reconnaissance," Hqs. Air Weather Service, undated, p. 4.

20. Paul A. Steves, letter to the author, 9 April 1994.

21. Paul A. Steves, letter to the author, 17 January 1994.

22. R. Cecil Gentry, "Hurricane *Debbie* Modification Experiments, 1969," *Science,* April 1970, p. 473ff;

23. William Anderson, *Hurricane Hunters,* p. 76.

24. "Is Man Upsetting the Weather?" *U.S. News and World Report,* 11 November 1963, p. 48.

25. Ibid., p. 46.

26. E. T. Blair, "Nature's Meanest Mood, "*The Airman,* January 1969.

27. Cathy Willis, "Rich" Courtney, Frank Baillie, "Aerographers Mates Celebrate," Naval reconnaissance history prepared for the 69th reunion of the Naval Weather Service Association, 1993, p. 34.

28. Ibid., p. 47.

29. Ibid., p. 48.

30. "Taming the Hurricane," *Time,* 1 September, 1969, p. 47.

31. Ibid.

32. Ibid.

33. Jim Perkins, Paul McVickar, *Documentary of the 53rd Weather Reconnaissance Squadron "The Hurricane Hunters,"* statement by Dr. Robert Simpson, Director, National Hurricane Center, 1968-1973, p. 25.

34. Lester Nelson, letter to the author, 9 December 1994.

35. Perkins, McVickar, *Documentary...* p. 35.

36. Ibid., p. 36.

37. Fuller, *Thor's Legions,* p. 385.

38. Arnold Zimmerman, letter to the author, 5 February 1994.

39. John D. Horn, "WB-47 Weather Reconnaissance," *PACAF Flyer,* January-February 1964, pps. not numbered. Lt. Col. John D. Horn was commander of Detachment 2 of the 55th WRS.

40. "WB-47E Flies First Mission," *AWS Observer,* Scott AFB, Ill., July 1963, p. 1.

41. Thomas C. Robison, letter to the author, 7 October 1994.

42. Perkins, McVickar, *Documentary...,* p. 23, 24.

43. Gary F. Frey, "The Hurricane Hunters," *JOURNAL, American Aviation Historical Society,* Spring 1980, p. 46.

44. Andrew Sparks, "Hurricane Hunters of Hunter," *Atlanta Journal and Constitution,* 30 August 1964, Sunday magazine section.

45. "Storm War Veteran Joins Hurricane Hunter Outfit," *Savannah Evening Press,* 16 June 1964.

46. Ibid.

47. Magilavy, fax, Ibid.

48. Ibid.

49. Robison, "Bried Genealogy of all known WC-130s, 1993, pp. 2, 3.

50. M. Clyde Garrison, letter to the author, 16 January 1995.

51. Robison, "Brief Genealogy . . . , p. 3.

52. Ibid.

53. John Collens, letter to the author, 11 March 1994. Major General (Ret) Collens, Placerville, California, is president of the Air Weather Reconnaissance; Wagner, Ibid., p. 430.

54. Donald E. Schertz, letter to the author, 22 November 1994.

55. *Eielson Times*, Eielson AFB, Alaska, 4 November 1964, p. 1.

56. *Albuquerque Journal*, 12 December 1964; Robison, Ibid.,

57. "MATS B-57 Squadron," *The MATS Flyer*, March 1962, p. 27.

58. Paul Steves, letter to the author, 9 April 1994.

59. Paul McVickar, letter to the author, 18 January 1994.

60. Fuller, *Thor*... Ibid., p. 271.

61. Ibid., p. 287.

62. Robison, "Brief Geneology . . . , p. 3.

63. John Hug, taped interview, 4 April 1994.

64. Fuller, *Thor*... , p. 288-89.

65. William Anderson, *Hurricane Hunters*, p. 88. Although this book was written as a novel, it is based on fact. Anderson is a former weather pilot with over two thousand hours of weather experience.

66. Fuller, *Thor's Legions*, p. 355.

67. Robison, "Brief Genealogy . . . ," p. 4.

68. Ibid.

69. Ed Blair "Where the Action Is," *The Airman*, January 1969, p. 18.

70. Deborah Shapley, "Rainmaking: Rumored Use Over Laos Alarms Arms Experts, Scientists," *Science*, 16 June 1972, p. 1216ff.; Fuller, Ibid., p. 291ff.

71. Arnett S. Dennis, *Weather Modification by Cloud Seeding*, p. 5.

72. Ibid., p. 243.

73. Henry Kissinger, *Years of Upheaval*, p. 1165.

74. Robison, "Brief Chronology of Air Force Aerial Weather Reconnaissance," October 1994., p. 3; Fuller, *Thor's Legions*, p. 318.

75. Robison, "Brief Chronology . . . , p. 3.

76. Tom Robison, letter to the author, 16 January 1994.

77. Memo from David Magilavy to the author, n.d., Subject: Weather Reconnaissance System AN/AM-15; *The AN/AMQ-15*, Bendix-Boeing brochure, September 1958.

78. Paul McVickar, letter to the author, 18 January 1994, p. 2,3.

79. Robert H. Simpson, Herbert Riehl, *The Hurricane and Its Impact*, p. 21.

80. Robison, "Brief Geneaology . . . , p. 4.

81. Sgt. Jeff McGuines, "They're Always Chasing Snowstorms," *The National Weather Service Observer*, March 1972, pp. 4, 5.

82. Ibid.

83. R. D. Gilmore, letter to the author, 17 January 1995. Mr. Gilmore was a Navy flight meteorologist.

84. Robert Smith, letter to the author, 10 February 1994.

85. Col. Paul McVickar, letter to the author, 18 January 1994.

86. "Supercooled Fog Dissipation by Airborne Dry Ice Seeding," Department of the Air Force, Chapter 14, pp. 14-1, 14-2; 23 November 1984; Fact sheet, "Fog Floggers, Ltd.," furnished by Lt. Col. Lars Olausson, C-130 historian.

87. Ibid.

88. Ibid.

89. John W. Collens, Maj. Gen., USAF (RET), letter to the author, 11 March 1994; Howard E. Lysaker, Col. USAF, Commander 11th WRS (MAC), *Cold Fog Abatement, Final Report, FY 72,* Department of the Air Force, 1972, p. 16ff.

CHAPTER 7

1. Cathy Willis, "Rich" Courtney, and Frank Baillie, "Aerographers Mates Celebrate," a history of Navy weather, prepared for the 69th anniversary of the Naval Weather Service Association, 1993, p. 4.

2. "Weather Satellite," *The Concise Columbia Encyclopedia, Third Edition,* p. 939.

3. John Fuller, *Thor's Legions,* p. 353-354.

4. William V. Yelton, personal letter to the author, 26 January 1995.

5. Military personnel with flight ratings received "flight pay" if they flew at least four hours each month. When there were few missions, flight crews, and desk-bound administrative personnel, often took "training" flights to qualify for flight pay. Thus "boring holes in the sky" was a term for flying useless missions.

6. Yelton, letter.

7. George Horn, *WC-135 Fact Sheet,* n.d., (c. July 1982).

8. "Commander's Comments," *Recon,* 9th Weather Wing, August 1967, pp. 1, 2.

9. Patrick Hughes, *Century of Weather Service,* p. 212.

10. Les Nelson, letter to the author, 9 December 1994.

11. "First Solo Assault on the Pole," *The National Geographic Magazine,* September, 1978, pp. 302-303.

12. Horn, *WC-135 Fact Sheet,* p. 3.

13. Al Lloyd, "Versatility Unlimited—the Boeing KC-135 Story," *Air International,* November 1980, p. 220.

14. Nelson, letter.

15. Thomas Robison, *Brief Chronology of Air Force Aerial Weather Reconnaissance,* p. 4.

16. Ralph L. Morgan, letter to the author, 5 September 1994.

17. Ibid.

18. Ed Mendell, "McClellan Crew in Loges Again—Apollo Launch," *Sacramento Union,* 15 April 1972, p. A-1

19. National Aeronautics and Space Administration, "Space, the New Frontier," quoted in *The Air Weather Service Observer,* December 1959, p. 8.

20. Fuller, *Thor's Legions,* p. 362.

21. Paul E. McVickar, letter to the author, 18 January 1994.

22. Ibid.

23. Ibid.

24. Robert Styger, letter to the author, 2 August 1994.

25. From the files of Don Cruse, Cdr., USN (Ret)

26. H. J. "Walt" Walter, *The Wind Chasers,* p. 5.

27. Ibid., pp. 151-159.

28. "Wind Chasers" was the title used for Navy reconnaissance crews by H. J. Walter, in his book by the same name.

29. Ibid., preface.

30. McVickar, letter.

31. Horn, Fact Sheet, p. 3.

32. Rita Markus, Nicholas Halbeisen, John Fuller, *Air Weather Service: Our Heritage-1937-1987,* p. 37.

33. In military service there are "active" and "reserve" personnel. Reserves are inactive and, technically, civilians, except when called to active duty for a special mission or regular "weekend" training for which they are paid. Thus hurricane recon "reserves" may hold regular civilian jobs except when called to active duty when hurricane missions demand their services.

34. *Biloxi* (Mississippi) *Sun Herald,* 9 June 1976.

35. John W. Pavone, Chief Aerial Reconnaissance Coordinator, All Hurricanes, letter to the author, 7 September 1994.

36. NCO—non commissioned officer.

37. Yelton, letter.

38. Ibid.

39. Parvone, Ibid.

40. Ibid.

41. Maj. Gen. John Collens, letter to the author, 10 January 1994.

42. McVickar, letter.

43. Arthur McCartan, letter to the author, 6 February 1994.

44. McVickar, Ibid.

45. "Soviets may have fired lasers at U.S. planes," *The Sacramento Bee—Final,* 10 November 1998, p. A10.

46. "Weather Reconnaissance Update," letter from Department of the Air Force, Hq. 7th Weather Wing (MAC), 14 September 1989.

47. John Ginovsky, "AWS unit flying into eye of technology," *Air Force Times,* 29 August 1988, p. 28.

48. Collens, letter.

49. Maj. Jim Perkins, "Documentary on the 53rd Weather Reconnaissance Squadron 'The Hurricane Hunters,'" script for a television production, produced by Lt. Col. Paul McVickar, undated, (c. 1990).

50. Jacquelyn Swearingen, "*Gabrielle* helped demonstrate value of U.S. 'hurricane hunter' flights," *The Miami Herald,* 9 September 1989, p. 12A.

51. Robert C. Sheets, "The National Hurricane Center—Past, Present and Future," *Weather and Forecasting,* June 1990, p. 216.

52. "Statement by Dr. Robert Sheets, director, National Hurricane Center," *NOAA'S Response to Weather Hazards — Has Nature Gone Mad?* in a hearing before the Subcommittee on Space, One Hundred Third Congress, 14 September 1993, p. 34-35.

53. Ibid.

54. Sheets, "Weather and Forecasting," p. 188-189.

55. Eugene Linden, "Burned by Warming," *Time,* 14 March 1994, p. 79.

56. Joe West, "AF tries again to cut back 'Storm Trackers'" *Air Force Times,* 18 May, 1992.

57. Ibid.

58. Congressman Ralph M. Hall, chairman of the Sub-Committee on Space of the Committee on Science, Space and Technology, *NOAA's Response . . .* p. 1.

59. Sheets, *NOAA's Response,* p. 35.

60. J. Madeline Nash, "Hurricane Onslaught," *Time,* 11 September 1995, p. 65.

61. Ibid., pp. 65, 66.

62. US Census Bureau, National Climatic Data Center, Associated Press research, "Waiting for the big one", *Greenville,* (Texas) *Herald Banner,* 15 July 1995.

63. Tony Bartelme, "Crews gather to trace storm," *The Post and Courier,* Charleston, South Carolina, 25 August 1992; Roy Deatherage, personal correspondence with the author, 10 October 1994.

64. Mark Patiky, "Factors that cause killer hurricanes," *Professional Pilot,* July 1933, p. 74. It is almost impossible to get consistent figures on storm damage which are usually totaled up years later.

65. Ibid, p. 42.

66. Ibid.

67. Gregory W. Withee, Acting Assistant Administrator for Satellite Information Services, National Oceanic and Atmospheric Administration, *Modernization of the National Weather Service,* hearing before the Subcommittee on Space, One Hundred Third Congress, 11 March 1993, p. 65.

68. Sheets, *NOAA's Response,* pp. 56, 65.

69. Elbert W. Friday, Jr., Assistant Administrator for Weather Services, National Oceanic and Atmospheric Administration, *Modernization of the National Weather Service,* Hearing before the Subcommittee on Space, One Hundred Third Congress, 11 March 1993, p. 17.

70. Withee, *Modernization . . . ,* p. 65.

APPENDIX

1. Chronology and lineage sketches have been compiled from one or more of the following: Historical Division, Air Weather Service, "Special Historical Study: Tracing of Organizational Structures, Weather Reconnaissance Squadrons, End of W.W.II to Present," 13 February 1959. Supplementing this study was later chronology from "Lineage Sketches, AWS Weather Reconnaissance Units," Air Weather Service, n.d.; The most complete lineages were found in Rita M. Markus, Nicholas F. Halbeisen, John F. Fuller, *Air Weather Service, Our Heritage, 1937-1987,* p. 140ff. Charles A. Ravenstein, The Albert F. Simpson Historical Research Center (USAF), *USAF Unit Lineage and Honors,* 15 July 1982; *Lineage Sketches,* of AWS Weather Reconnaissance Units, non-dated document; Office of Information News Release, 53rd Weather Reconnaissance Squadron (MAC), not dated. There are slight differences in dates and names in these documents.

2. Documents in files of the author.

Bibliography

AAF, The Official World War II Guide to the Army Air Forces. New York: Bonanza Books, 1988.

Ainlay, Maj. George. "1944—Year of the Great Hurricane." *Science Digest,* May 1944, pp. 39-40.

Allaby, Michael. *Air—The Nature of Atmosphere and the Climate.* New York: Facts on File, Inc., 1992.

Anderson, William C. *Hurricane Hunters.* New York: Crown Publishers, Inc., 1975.

——. "Edward R. Murrow and the Eye of the Hurricane." *The Retired Officer Magazine,* April 1995, pp. 51-55.

Arnold, H. H. *Global Mission.* New York: Harper & Brothers, 1949.

Arnold, John K. "Assignment of Personnel, Special Orders 201.22." Regional Control Office, 9th Weather Region, Morrison Field, West Palm Beach, Florida, 28 December 1944.

Bates, Charles C. "Thoughts on D-Day Forecast." *Newsletter, D-Day Commemorative Edition,* Air Weather Association, June 1944, p. 12.

—— and John Fuller. *America's Weather Warriors, 1814-1985.* College Station Texas: Texas A&M University Press, 1986.

Blair, Ed. "Where the Action Is." *The Airman,* January 1969, pp. 15-18.

——. "Nature's Meanest Mood." *The Airman,* January 1969, pp. 16-21.

Blake, Robert W. "Nightmare Over China." *International Air Review,* Winter 1951, pp. 50-59.

Bliss, Edward, Jr., editor. *In Search of Light, The Broadcasts of Edward R. Murrow 1938-1961.* New York: Alfred A. Knopf, 1967.

Boeing Company. "Background Information—Boeing B-29 Superfortress." News release, 13 January 1971.

Bowers, Neal M. "Bikini Atoll." *The World Book Encyclopedia.* Chicago: Field Enterprises Educational Corporation, 1959.

Bradley, Omar N. *A Soldier's Story.* New York: Rand McNally & Company, 1951.

Brown, Andrew H. "Men Against the Hurricane." *National Geographic Magazine,* October 1950, pp. 357ff.

Brown, Ewing Franklin. *The Weathermen Let Them Fly.* Eighth Air Force Historical Society, 1993.

Bundgaard, Robert C. "D-Day Forecast Fifty Years Later." *Newsletter, D-Day Commemorative Edition,* Air Weather Association, June 1944, p. 7

Burgmeier, John J. III. "Orphans of the Storm, Part One." *Air Classics,* April 1986, pp. 26-33.

———. "Orphans of the Storm, Part Two." *Air Classics,* May 1986, pp. 23-26, 56-62, 74.

Burpee, Robert W. "Gordon E. Dunn: Preeminent Forecaster of Midlatitude Storms and Tropical Cyclones." *Weather and Forecasting,* American Meteorological Society, December 1989.

"Camille-A Calamity at 205 Miles an Hour." *U.S. News and World Report,* September 1, 1969, p. 35.

Carlisle, Norman. "Our Weather is Changing." *Coronet,* January 1954, pp. 17ff.

———, editor. *The Air Forces Reader,* New York: The Bobbs-Merrill Company, 1944.

Carroll, Lewis. *Walrus and the Carpenter. St.,* 1,b.

China-Burma-India, Theater History, 1941-1945. Microfilm Index 940, Roll No. A8155. Headquarters, United States Air Force Historical Research Center, Maxwell Air Force Base, Alabama.

Close, Winton R. "B-29s in the CBI—A Pilot's Account." *Aerospace Historian* 22 March 1983.

Clouse, Capt. Gale E., Jr. "Tracking the Storm—Typhoon Chasers of the Pacific." *The MAC Flyer,* June 1983, pp. 8-11.

Cookman, Aubry O. "Top of the World Weather Run." *Popular Mechanics,* November 1948, pp. 99-100.

Copp, DeWitt S. *Forged in Fire-Strategy and Decisions in the Air War over Europe.* Garden City: Doubleday & Company, Inc., 1982.

Cox, Capt. L. Scott. "Hunting for Trouble." *The MAC Flyer,* January 1978, pp. 13-15.

Craven, Wesley Frank, and James Lea Cate, editors. *Men and Planes.* Vol.

6 of *The Army Air Forces in World War II*. Chicago: University of Chicago Press, 1953.

——. *The Pacific: Matterhorn to Nagasaki*. Vol. 5 of *The Army Air Forces in World War II*. Chicago: University of Chicago Press, 1953.

——. *Services Around the World*, Vol. 7 of *The Army Air Forces in World War II*. Washington, D. C.: United States Government Printing Office, 1983.

"Curbing Hurricanes—the Chances." *U.S. News and World Report*, September 1, 1969, pp. 34-36.

Dennis, Arnett S. *Weather Modification by Cloud Seeding*. New York: Academic Press, 1980.

"Destructive Weather." *Texas Almanac-1994-1995*. Dallas: *The Dallas Morning News*, 1993.

Diamond, Nina L. "Tropical Delight—Finding Cures Among the Wreckage of a Hurricane." *Omni*, May 1994, p. 22.

——."Out of the Doldrums." *Time*, 25 September 1944, p. 61.

Dorsey, H. G., Jr. and Oscar Shaftel. "Sky Quirks." *Air Force*, No. 6, June 1944, p. 31.

Dotson, Cecil. "How D-Day Was Chosen." *Newsletter, D-Day Commemorative Edition*, Air Weather Association, June 1994, pp. 2-7.

——. "D-Day and the Weather." *Newsletter, D-Day Commemorative Edition*, Air Weather Association, June 1944, p. 15.

Duckworth, Joseph P. Letter to F. W. Reichelderfer, Chief, Weather Bureau, United States Department of Commerce, Washington, D.C., 13 September 1943.

Dunn, Gordon E. and Banner I. Miller. *Atlantic Hurricanes*. Baton Rouge: Louisiana State University Press, 1960.

Earney, Capt. Lyle P. "At War and Peace-The B-29 Superfort Served Well." *Tropic Topics*, (Guam), 25 January 1957, pp. 4-5.

Eisenhower, David. *Eisenhower At War 1943-1945*. New York: Random House, 1986.

Eisenhower, Dwight D. *Crusade in Europe*. New York: Doubleday & Company Inc., 1948.

Fowle, Barry E. "The Normandy Landing." *Army History, The Professional Bulletin on Army History*, Washington, D. C., Spring 1994, p. 1.

Frey, Gary F. "The Hurricane Hunters." *Journal*, American Aviation Historical Society, Spring 1980, pp. 45-56.

Friends Bulletin. Air Force Museum Foundation, Inc., Summer/Fall, 1986.

Fuller, John F. *Thor's Legions, Weather Support to the U.S. Air Force and Army*. Boston: American Meteorological Society, 1990.

——. *Thor's Legions*. Unpublished Manuscript,

——. "A Lesson from History . . . the WB-50D." *Friends Bulletin*, United Air Force Museum, Summer/Fall 1986, pp. 18-23

Gann, Ernest K. *Fate is the Hunter.* New York: Simon & Schuster, 1961.

Gentry, R. Cecil. "Hurricane Debbie Modification Experiments, August 1969." *Science,* 24 April 1970, pp. 473-475.

Giles, Brian D., editor. *Meteorology and World War II.* Birmingham, Great Britain: Royal Meteorological Society, Department of Geography, University of Birmingham, 1987.

Ginovsky, John. "AWS unit flying into eye of technology." *Air Force Times,* 29 August 1988, p. 28.

Gordon, Arthur. "The Conquest of Fear." *Air Force,* June 1944, p. 11

"Greenhouse Fact Sheet." Defense Nuclear Agency, Washington, D.C., not dated.

Gulliver, Art. "Langford Lodge, North Ireland." *Newsletter,* Air Weather Reconnaissance Association, July 1993, p. 4.

——. "Who Told the Weather?" *Journal* of the 7th Photo Recon Group, September 1992.

Gurney, Gene. *Journey of the Giants.* New York: Cowar-McCann, 1961.

[Guthland, Robert]. "'Tai-Fung' Chasers. *Flying Safety,* November 1956.

Hailey, Foster. "This is the Superfortress." In *The Air Forces Reader,* edited by Norman Carlisle. Newark: University of Delaware Press, 1978.

Haugland, Vern. *The Eagle's War, The Saga of the Eagle Squadron Pilots 1940-1945.* New York: Jason Aronson, 1982.

Haynes, B. C. *Techniques of Observing the Weather.* New York: John Wiley & Sons, Inc., 1947.

Henry, Walter K., Dennis M. Driscoll and J. Patrick McCormack. *Hurricanes on the Texas Coast.* College Station Texas: Center for Applied Geosciences, Texas A&M University, July 1975.

Henderson, Rodney S. "USAF Aerial Weather Reconnaissance Using the Lockheed WC-130 Aircraft." *Bulletin American Meteorological Society,* September 1978, pp. 1136-1143.

Holy Bible, Genesis 8:8-12. KJV.

Howes, Lewis L. *The Nuclear Explosion Detection System, and the Role of the Air Weather Reconnaissance Program.* Private publication, not dated (c. 1944).

Hughes, Patrick. *A Century of Weather Service, A History of the Birth and Growth of the National Weather Service.* New York: Gordon and Brech, Science Publishers, Inc., 1970.

"The Hurricane Hunters—Weathermen at Work." *Aerospace Safety,* United States Air Force, December 1968, pp. 16-19.

"Hurricane hunters add to jet power." Greenville *(Tex) Herald Banner,* 10 December 1994.

"Hurricane watchers review busy season." *The Dallas Morning News,* 27 November 1995, p. 4-A.

Hurricane Weather Reconnaissance. Hearings before the Subcommittee on Natural Resources, Agriculture Research and Environment, of the Committee on Science, Space, and Technology, U.S. House of Representatives, One Hundred First Congress, April 7, 1989. Washington, D. C.: U.S. Government Printing Office, 1989.

"Ice Islands as Bases," *Science Newsletter.* 25 November 1950, p. 58.

"Ice Islands." *Time*, 27 November 1950, p. 66.

"Into a Hurricane's Eye." *Newsweek*, 25 September 1944, p. 45-46.

"Is Man Upsetting the Weather?" *U.S. News and World Report*, 11 November 1963, pp. 46-48.

Jones, L. S. and Edward J. Machala. *History of AAF Weather Reconnaissance Squadron (Test) No. 1, 21 August-15 December 1943.* Presque Isle, Maine, 4 March 1944.

Jordanoff, Assen. *Through the Overcast, The Weather and the Art of Instrument Flying.* New York: Funk & Wagnalls Company, 1938.

Kane, Joseph N. *Famous First Facts.* First Edition. New York: The H. W. Wilson Co., 1933.

Karig, Walter, Russell L. Harris and Frank A. Manson. *Battle Report*, Victory in the Pacific. New York: Rinehart and Company, Inc., 1949.

Kemp, Capt. George. *Special Orders, No. 112.* Headquarters, Columbia Army Air Base, Columbia, South Carolina, 21 April 1944.

Kissinger, Henry. *Years of Upheaval.* Boston: Little, Brown and Company, 1982.

Krick, Irving P. and Lee Edson. "How D-Day Was Chosen." *Newsletter, D-Day Commemorative Edition*, Air Weather Association, 1 June 1944, p. 2

Larrabee, Eric. *Commander in Chief: Franklin Delano Roosevelt, His Lieutenants and Their War.* New York: Harper & Row, 1987.

Linden, Eugene. "Burned by Warming." *Time*, 14 March 1994, p. 79.

Lockhart, Gary. *The Weather Companion.* New York: John Wiley & Sons, Inc., 1988.

Lloyd, Al. "Visibility Unlimited—The Boeing KC-135 Story." *Air International*, November 1980. pp. 220-238.

Ludlum, David M. *Early American Hurricanes, 1492-1870.* Boston: American Meteorological Society, 1963.

"Making Rain While the Sun Shines." *Science News*, 27 July 1985, p. 88.

Magilavy, David, editor. *Newsletter*, Air Weather Reconnaissance Association, 1 May 1994.

Markham, Charles G. "A Balloon Blower in Wartime China." *Newsletter, D-Day Commemorative Edition*, Air Weather Association, June 1944, p.

——. "Spit in the Typhoon's Eye." *The Retired Officer*, May 1974, pp. 18-21.

Markus, Rita M., Nicholas F. Halbeisen and John Fuller. *Air Weather Ser-*

vice: Our Heritage 1937-1987. Scott AFB, Illinois: Military Airlift Command, United States Air Force, 1987.

Matt, John. Crewdog-A Saga of a Young American. Hamilton, Virginia: Waterford Books, 1992.

McCormick, Elsie. "They Hunt for Bad Weather." Saturday Evening Post, 30 March 1946, p. 17, 102-104.

McCullough, David. Truman. New York: Simon & Schuster, 1992.

McGuiness, Sgt. Jeff. "They're always chasing snowstorms" The Air Weather Service Observer, March 1972, pp. 5-5.

Misner, Edward J. "St. Elmo's Fire." Encyclopaedia Britannica, Chicago: Encyclopaedia Britannica, Inc., 1954, Vol. 1, 1954.

Modernization of the National Weather Service. Hearing before the Subcommittee on Space of the Committee on Science, Space, and Technology, U.S. House of Representatives, One Hundred Third Congress, March 11, 1993. Washington D. C.: U.S. Government Printing Office, 1993.

Murchie, Guy. Song of the Sky. Boston: Houghton Mifflin Company, 1954.

Nalivkin, D. V. Hurricanes, Storms and Tornadoes. Rotterdam: A. A. Balkema, 1983.

"Naming of Hurricanes." Bulletin of the U.S. Department of Commerce, National Weather Service, undated (c. 1990)

"NASA Weather Prophecy." The Air Weather Service Observer, December 1959, p. 8.

Nelan, Bruce W. "Ike's Invasion." Time, 6 June, 1994, pp. 36-49.

Nichols, Bruce. "Cyclone forecast on way." The Dallas Morning News, 1 June 1986.

NOAA's Response to Weather Hazards—Has Nature Gone Mad? Hearing before the Subcommittee on Space of the Committee on Science, Space, and Technology, U. S. House of Representatives, One Hundred Third Congress, 14 September 1993. Washington, D. C.: U.S. Government Printing Office, 1993.

"Notes on the Naming of Hurricanes." U.S. Department of Commerce-1956. 23 December 1955, p. 2.

Nye, Bill. "Big Energy in Thin Air." Omni, May 1994, p. 13,

Orville, Howard T. "Stand by for Climate Control." Saturday Evening Post, Nov. 29, 1958, pp. 19-21.

"Pacifying Ginger. Time, October 11 1971, p. 61.

Patiky, Mark. "Factors that Cause Killer Hurricanes." Professional Pilot, July 1993, pp. 75-78.

Perkins, Maj. Jim and Lt. Col. Paul McVikar. "Documentary on the 53rd Weather Reconnaissance Squadron, 'The Hurricane Hunters.'" Undated, (c. 1990).

Phillips, Kathryn. "Breaking the Storm." *Discover, The World of Science.* May 1992, pp. 2ff.

Phillips, Ralph N. "Navigation Problems in England." *Air Force,* 6 June 1944, p. 36-37.

Posey, Carl. "Hurricanes-Reaping the Whirlwind." *Omni,* March 1994, pp. 34-47.

Purrett, Louise. "How to Subdue a Hurricane." *Science News,* pp. 128-129.

Quinn, Chick Marrs. *The Aluminum Trail.* Lake City, Florida: The Sequin Press, 1989.

Rackliff, P.G. Brian Giles, editor. "Meteorological Reconnaissance Flights." *Meteorology in W.W.II."* Birmingham (England): Royal Meteorological Society, Department of Geography, University of Birmingham, 1987.

Reiter, Elmar R. *Jet Streams-How Do They Affect Our Weather?* Garden City, New York: Doubleday & Company, Inc., 1967.

———. *Jet-Stream Meteorology.* Chicago: The University of Chicago Press, 1961.

Robison, Thomas C. "Brief Chronology of all Known WC-130's." Personal publication, 6 September 1994. pp. 1-5.

———. "Brief Chronology of Air Force Aerial Weather Reconnaissance. " Personal publication, October 1994, pp. 1-6.

Roosevelt, Franklin D. Address, 15 March 1941.

Sadler, Bill. "Hunt the Hurricane." *The MAC Flyer,* June 1980, pp. 10-11.

———. "Now They Wear Wings." *The MAC Flyer,* December 1970, pp. 18-19.

Sevier, M. L. *Fact Book-Aerial Sampling and Weather Reconnaissance.* Headquarters 41st Rescue and Weather Reconnaissance Wing (MAC), Department of the Air Force, McClellan Air Force Base, California, November 1983.

Senter, William Oscar. *Special Orders Number 118.* Headquarters, Weather Wing, Asheville, North Carolina, 18 May 1944.

Sheets, Robert. "The National Hurricane Center-Past—Present and Future." *Weather and Forecasting,* American Meteorological Society, June 1990, p. 200.

Simpson, Robert H. and Herbert Riehl. *The Hurricane and Its Impact.* Baton Rouge: Louisiana State University Press, 1981.

Sosin, Milton. "Daily News Writer Hunts Hurricane in 100 MPH Winds." *Miami Daily News,* 13 September 1944.

Sparks, Andrew. "Hurricane Hunters of Hunter." *The Atlanta Journal and Constitution Magazine,* 30 August 1964.

Special Historical Study: Tracing of Organizational Structures, Weather Reconnaissance Squadrons, End of W.W.II to Present. Historical Division, Headquarters, Air Weather Service, February 1959.

Spencer, Otha C. *Flying the Hump.* College Station: Texas A&M University Press, 1992.

Stilwell, Joseph W. *The Stilwell Papers.* Edited and arranged by Theodore H. White. New York: William Sloan Associates, 1948.

"Storm stalkers get jump on hurricanes." *Business Week,* 15 July, 1967. pp. 106-108.

"Storm War Veteran Joins Hurricane Hunter Outfit." *Savannah Evening Press,* 15 June 1964.

Summary of 17th Weather Squadron, World War II History. Microfilm, Air Force Historical Research Center, Maxwell Air Force Base, Alabama.

Sumner, H. C. "North Atlantic Hurricanes and Tropical Disturbances of 1944." *Monthly Weather Review,* December 1944, p. 340.

Swarts, Jim. "Orphans of the Storm." *Stars and Stripes, 1944.*

——. Jim. "Tracing the Organizational Structures, Weather Reconnaissance Squadrons, End of W.W.II to Present." February 1959, pp. 1-2.

"Taming the Hurricane," *Time,* 1 September 1969, p. 47-48.

Tannehill, Ivan Ray. *Hurricanes, Their Nature and History.* Ninth Revised Edition. New Jersey: Princeton University Press, 1950.

——. *The Hurricane Hunters.* New York: Dodd, Mead & Company, 1963.

——. *Weather Around the World.* New Jersey: Princeton University Press, 1952.

Taylor, Capt. Elden C. *History of the 53rd Weather Reconnaissance Squadron.* Ramey Air Force Base, Puerto Rico, 1 January 1969 to 30 June 1969.

"Thunder Over the North Atlantic," *Fortune,* November 1944, pp. 153-206.

Townsend, Jeff. *Making Rain in America: A History.* Lubbock, Texas: International Center for Arid and Semi-Arid Land Studies, Texas Technological University, 1975.

Trewartha, Glenn T. *The Earth's Problem Climates.* Second Edition. Madison: The University of Wisconsin Press, 1981.

Truman, Harry S. *Memoirs, Vol. One-Year of Decisions.* Garden City, New York: Doubleday & Company, Inc. 1955.

Tuchman, Barbara W. *Stilwell and the American Experience in China, 1911-1945.* New York: The MacMillan Company, 1970.

Tunner, William H. *Over the Hump.* Washington, D.C.: Office of Air Force History, United States Air Force, 1985.

Uemura, Naomi. "First Solo Assault on the Pole." *National Geographic Magazine,* September 1978, pp. 298-325.

United States Air Force Museum. *Friends Bulletin,* Summer/Fall 1986, Vol. 9. Published by The Air Force Museum Foundation, Inc.

United States Joint Chiefs of Staff, "AC/AS OC&R - AFRWX," 26 May 1944, signed by Arnold and Marshall.

Wagner, Ray. *American Combat Planes.* Garden City, New York: Doubleday & Company, 1982.

Walter, H. J. *The Wind Chasers-The History of the U.S. Navy's Atlantic Fleet 'Hurricane Hunters',* " Dallas: Taylor Publishing Co., 1992, 169 p.

Watson, Bruce. "Jaysho, moasi, dibeh, ayeshi, hasclishnih, beshlo, shush, gini-War and Remembrance: Navajo Code Talkers," *Smithsonian,* August 1993, pp. 34-40+.

——. "Code of the Navajos." *Reader's Digest,* December 1993, pp. 39-40+.

"Weather Satellite." *The Concise Columbia Encyclopedia,* Third Edition. New York: Columbia University Press, 1994.

Wiggin, Bernard L. "A Weatherman has been Called the Most Useful Man of his Age." *Smithsonian,* December 1970, pp. 50-51.

West, Joe. "AF tries to cut back 'Storm Trackers.'" *Air Force Times,* May 18, 1992.

Willis, Cathy, "Rich" Courtney and Frank Baillie. *Aerographers Mates Celebrate-Sixty-Nine Years of Environmental Support.* Naval Weather Service Association, 1993.

Wolfe, Virginia. *Mrs. Galloway,* 1925.

Wolkomir, Richard. "Electric Sky." *Omni,* March 1994, pp. 50-54.

"Yates Necrology." *Bulletin of the American Meteorological Society,* June 1944.

Young, William. "The Meteorological Air Observer's Standpoint." *Newsletter,* Air Weather Reconnaissance Association, July 1994, p. 4.

Zimmerman, Kent. "Air Weather Service." *Personal Memoirs.* Unpublished manuscript, p. 204.

Index

Flying the Weather was composed into type on a Linotronic 300 Imagesetter, using a Macintosh IIx computer and Adobe PageMaker v6.0 software. The body type is ten point Trump Mediaeval with two points of space between the lines. Trump Mediaeval was also selected for display. The book was typeset at East Texas State University Instructional Printing, printed offset and bound by Henington Publishing. The paper on which this book is printed carries acid-free characteristics for an effective life of at least three hundred years.

THE COUNTRY STUDIO: CAMPBELL, TEXAS

Order Form

Please send me _____ copies of Flying the Weather
at $14.95 each.

Subtotal _____

Texas residents add 7.25% sales tax _____

Shipping/handling charge $ 3.00

Total due _____

Order from: Ship to:

The Country Studio   _____

Route 2, Box 54   _____

Campbell, Texas 75422   _____

- CLIP- -

Order Form

Please send me _____ copies of Flying the Hump
at $14.95 each.

Subtotal _____

Texas residents add 8.25% sales tax _____

Shipping/handling charge $ 3.00

Total due _____

Order from: Ship to:

Texas A&M
University Press

Drawer C
College Station, Texas 77843-4354
or call toll-free: 1-800-826-8911
FAX: 409-847-8752